Radar Meteorology

Radar is a key instrument used in meteorology for monitoring wind and precipitation, and has become the primary tool used for short-term weather forecasting. This practical textbook introduces the fundamental concepts behind radar measurements and their meteorological interpretation.

The first part of the book provides the essential background theory behind radar measurements to guide students and practitioners in the proper interpretation of radar reflectivity, Doppler velocity, and dual-polarization imagery. Operational applications are then explored, such as how radar imagery can be used to analyze and forecast convective and widespread weather systems. The book concludes with an overview of current research topics, including the study of clouds and precipitation using ground-based and spaceborne radars, signal processing, and data assimilation.

Numerous full-color illustrations are included, as well as case studies, and a variety of supplementary electronic material, including problem sets and animated time sequences of images to help convey complex concepts. This book is a valuable resource for advanced undergraduate and graduate students in radar meteorology and other related courses, such as precipitation microphysics and dynamics. It will also make a useful reference for researchers, professional meteorologists, and hydrologists.

Frédéric Fabry is an Associate Professor at McGill University in Montreal, Canada, where he teaches topics in radar, meteorology, and the environment. He is also the Director of the Marshall Radar Observatory at McGill. His research covers various facets of radar meteorology, from technical aspects such as signal processing to applications of radar in hydrology and in numerical weather modeling, including more traditional radar meteorology research such as the characterization of the melting layer of precipitation. He was awarded the 2004 Canadian Meteorological and Oceanographic Society President Prize for the development of a technique to estimate the refractive index of air using ground targets.

Radar Meteorology

Principles and Practice

FRÉDÉRIC FABRY

McGill University

CAMBRIDGE
UNIVERSITY PRESS

CAMBRIDGE
UNIVERSITY PRESS

University Printing House, Cambridge CB2 8BS, United Kingdom

One Liberty Plaza, 20th Floor, New York, NY 10006, USA

477 Williamstown Road, Port Melbourne, VIC 3207, Australia

4843/24, 2nd Floor, Ansari Road, Daryaganj, Delhi - 110002, India

79 Anson Road, #06-04/06, Singapore 079906

Cambridge University Press is part of the University of Cambridge.

It furthers the University's mission by disseminating knowledge in the pursuit of education, learning and research at the highest international levels of excellence.

www.cambridge.org
Information on this title: www.cambridge.org/9781108460392

First published 2015
First paperback edition 2017

A catalogue record for this publication is available from the British Library

Library of Congress Cataloging in Publication data
Fabry, Frédéric, 1967–
Radar meteorology : principles and practice / Frederic Fabry, McGill University.
pages cm
Includes bibliographical references and index.
ISBN 978-1-107-07046-2
1. Radar meteorology. 2. Radar. 3. Meteorology. I. Title.
QC973.5.F33 2015
551.63´53–dc23
2015008279

ISBN 978-1-107-07046-2 Hardback
ISBN 978-1-108-46039-2 Paperback

Additional resources for this publication at www.cambridge.org/fabry

To Nicole, Roland, and Véronique

Contents

Preface

"Radar meteorology" is an odd specialty in atmospheric science. As opposed to synoptic meteorology or cloud physics, its focus has been on the instrument from the start. Radar enabled us to observe and understand many previously unknown phenomena. What we could do with this wonderful tool drew together a vibrant community of researchers whose main point in common was the use or the development of radars for meteorology. The instrument became the center of this community. As a result, when radar meteorologists meet, many often talk about instrument characteristics such as frequency, beam width, and transmit power before they talk of science. Early influential textbooks reflected that state of affairs, and many current introductory ones follow the same mold: they tend to be very radar focused, even those that do not describe in great detail the radar and its workings, and they are not very application oriented, despite the fact that the very reason we use radars is for what it allows us to see and do, from meteorological studies to short-term forecasting. Introductions to satellite meteorology, another technically oriented specialty, have managed to free themselves from their heritage: textbooks on how to use satellite imagery can be found, as well as more traditional books focused on radiative transfer. But somehow, introductions to radar meteorology have failed to do so.

Yet, the average weather radar user has changed. Radar is an operational instrument in many countries. It offers the forecasters what is generally the best opportunity to detect rapidly developing storms, and the last one to evaluate whether the weather evolves as expected or not. Also, the twenty-first-century researcher has a different focus: while efforts to improve and better understand radar data remain the principal objective of a necessary core of specialists, the emphasis has shifted toward meteorology and how to make the best use of the rich information provided by radar. Traditional books introducing radar meteorology, and the courses that rely on them, have gradually become disconnected from this changing reality. Furthermore, the historical focus of radar meteorology on the instrument and the physics of its measurements has made the subject difficult to teach and very dry to nonspecialists. This is regrettable given how wonderfully radar can be used to illustrate and understand weather phenomena, and to reinforce the learning of meteorological concepts seen in other courses, especially in precipitation microphysics and dynamics. As a result, a proper introduction on the use of radar for meteorology is often lacking in many university or professional programs, an illogical outcome given the current use of radar in operations and research. The frustration I experienced in the way radar meteorology was (not) taught, including to those who would not take up a career based on radar research, pushed me to write this introductory book.

To be fair, there is one good reason why many textbooks emphasize the more technical side of the basic principles of radar measurements: a proper interpretation of the imagery observed and how it may be corrupted can only be achieved after a thorough understanding of exactly what quantities a radar measures and how it does so. The day when radar imagery will be uncorrupted, free of all unwanted elements, and unambiguous has yet to arrive. Hence, one must still understand how the measurements are taken, what can corrupt them, and how to recognize what is correct from what is suspect. The book must still begin with a sufficient description of the basic principles of radar detection capability for the reader to be able to understand the nature and peculiarities of the data and how they can be contaminated. Subsequent chapters can then concentrate on the uses of radar data, starting with operational uses and gradually shifting to research applications.

An appendix introducing key mathematical and statistical concepts used in radar data analysis and processing completes the book. There are a few ways to use it depending on the level of the course and its emphasis. The appendix can be read as a block after Chapter 3 in the context of a thorough graduate-level course, or by subsections on a need basis, as these subsections are being referred to in the main text. It could also be skipped if the focus of the course is on the operational uses of radar.

The book chapters can be read in a different order depending on one's interest and focus. The cover-to-cover approach works well as an introduction for future researchers with a meteorology focus. A reader interested in how to use radar data operationally should focus on Chapters 1–8, continuing up to Chapter 12 if the first chapters stimulated your curiosity. A more traditional order of topics of radar meteorology starting with instrument and theory first would be Chapters 1–3, Appendix A, Chapter 13, and then 4–12, possibly skipping over Chapters 7 and 8.

This book has been made possible thanks to the contributions of many. Tony Banister, Don Burgess, and WenChau Lee offered background documents. Images and data were provided by Wayne Angevine, Aldo Bellon, William Brown, George Bryan, Guy Delrieu, Marielle Gosset, Robin Hogan, Robert Houze, Paul Joe, Sigrún Karlsdóttir, Jennifer Kay, Alamelu Kilambi, Pavlos Kollias, Witold Krajewski, Matthew Kumjian, Paul Markowski, Véronique Meunier, Kenji Nakamura, Rita Roberts, Steve Rutledge, Alan Seed, Matthias Steiner, Madalina Surcel, Pierre Tabary, Roger Wakimoto, and Isztar Zawadzki, in addition to the American Meteorological Society, Cambridge University Press, Elsevier, Environment Canada, the Institution of Engineering and Technology, InTech, NASA, NCDC, NOAA, Prosensing, Springer, Selex, the University Corporation for Atmospheric Research, and Wiley. Alexandra Anderson-Frey, Aldo Bellon, Alexandra Cournoyer, Véronique Meunier, and Pierre Vaillancourt provided valuable help and feedback on the manuscript, while regular discussions with Isztar Zawadzki shaped the contents of several chapters. At Cambridge University Press, Emma Kiddle, Rosina Piovani, Jonathan Ratcliffe, and Zoë Pruce shepherded the book project. Last but not least, thanks to my family, who pushed me along this long project and supported my absences during its execution, and to everyone else who kept asking me "*when are you finishing your book?*": I am finally able to provide them an answer.

Notation

List of symbols

a	generic coefficient of a power-law relationship
a	real part of a generic complex number, Appendix A.5
$a_{1,2,j}$	real parts of the complex numbers z_1, z_2, z_j
a_e	radius of the Earth
A	amplitude of the radar signal
A_{HH}	signal amplitude obtained when transmitting and receiving at horizontal polarization, also known as the *copolar* amplitude at horizontal polarization
A_{HV}	signal amplitude obtained when transmitting at horizontal polarization and receiving at vertical polarization
A_{VH}	signal amplitude obtained when transmitting at vertical polarization and receiving at horizontal polarization, also known as the *cross-polar* amplitude of the signal
A_{VV}	signal amplitude obtained when transmitting and receiving at vertical polarization, also known as the *copolar* amplitude at vertical polarization
b	generic exponent of a power-law relationship
b	imaginary part of a generic complex number, Appendix A.5
$b_{1,2,j}$	imaginary parts of the complex numbers z_1, z_2, z_j
B	bandwidth of the receiver
c	speed of light
c_j	complex weights to bases functions of a Fourier series, Eq. (A.37)
c_p	specific heat capacity of air at constant pressure
c_s	speed of sound, Eq. (5.3)
c_v	specific heat capacity of air at constant volume
$\text{cov}_{X,Y}$	covariance between the samples X and Y, Eq. (A.13)
$\text{cov}_{X,Y}[l]$	covariance between the samples X and Y at lag l, Eqs. (A.15) and (A.24)
C_n^2	refractive index structure parameter, Eq. (2.9)
dBZ	reflectivity factors in units of decibels, Eq. (3.5)
D	diameter of hydrometeor or target
D_a	diameter of a parabolic reflector
e	partial pressure of water vapor
$E(\lambda)$	Flux of energy at wavelength λ
E_ω	Power spectrum value at wavenumber ω, Eq. (A.41)

f	transmit frequency of the radar
$f()$	function
f_{IF}	intermediate frequency, or frequency of the radar return signal after mixing
f_j	value of a generic discrete function at sample j, Eq. (A.40)
f_{LO}	frequency of the internal local oscillator
f_{lobes}	fraction of the integral of the gain belonging to the sidelobes
f_r	pulse repetition frequency, Eq. (2.14) and Fig. 2.14
$F(\omega)$	component of a Fourier transform at wavenumber ω, Eq. (A.39)
$\mathcal{F}()$	Fourier transform operation, Eq. (A.39)
F_ω	component of a discrete Fourier transform at wavenumber ω, Eq. (A.40)
g	acceleration of Earth's gravity
G	gain (or directivity) of the antenna, Eqs. (13.1) and (13.2)
H_o	scale height
H_j	sample j of the received signal time series at horizontal polarization
i	$\sqrt{-1}$
i	unit vector in the x (east–west) direction, pointing east
I	component of the received signal in phase with the reference signal, Eq. (A.22)
I_j	component of the sample j of the received signal time series in phase with the reference signal
I_{S_j}	component of the sample j of the received signal time series originating from targets in phase with the reference signal
j, j_1, j_2	indices of series of values
j	unit vector in the y (north–south) direction, pointing north
k	Boltzmann constant, Eq. (13.3) only
k	index to drop or target number
k	unit vector in the z (up–down) direction, pointing up
k_e	multiplier to Earth radius for the k_e Earth radius approximation, Eq. (2.13)
$\|K^2\|$	dielectric constant of the scatterers, Eq. (3.3), Table 3.1
K_{dp}	specific differential propagation phase delay
$\|K_w^2\|$	dielectric constant of liquid water
l_p	longitude of the radar pulse, Eq. (A.1b)
l	lag, or offset in sample number
l_{rad}	longitude of radar
L	generic horizontal distance
L	interval over which a function f is assumed periodic, Appendix A.5
LDR	linear depolarization ratio, Section 6.2.2
L_p	latitude of the radar pulse, Eq. (A.1a)
L_{rad}	latitude of radar
M	number of measurements in a sample
M_s	number of measurements or subsamples in a sample
M_λ	number of wavelengths over which a measurement of I and Q is made
n	refractive index of air, Eq. (2.8)
$n(\lambda)$	complex refractive index of a target

n_1, n_2	refractive index of medium 1 and 2, Fig. 2.9
n_p	number of transmit pulses
N	refractivity of air, Section 2.3.1
N	Fraction of the received signal originating from noise, Appendix A.5
$N(D)$	number of scatterers of diameter D per unit volume
$N_f(D)$	number of scatterers of diameter D per unit volume after the growth process, Eq. (9.2)
$N_i(D)$	initial number of scatterers of diameter D per unit volume before the growth process
N_o	number of scatterers of diameter 0 per unit volume in the context of an exponential drop size distribution
$p()$	probability function
P	air pressure
P_d	power from the direct echo in the context of the mirror image technique, Fig. 12.6
P_g	power from the ground or sea surface in the context of the mirror image technique, Fig. 12.6
P_{HH}	signal power received at horizontal polarization given a transmission at horizontal polarization
P_m	power from the mirror image in the context of the mirror image technique, Fig. 12.6
P_N	noise power, Eq. (13.3)
P_r	returned or received power, Eq. (3.2)
$\overline{P_r}$	average returned or received power
P_t	power of the transmit pulse
P_{VV}	signal power received at vertical polarization given a transmission at vertical polarization
q	exponent of the power spectrum of atmospheric patterns, Section A.4.4
Q	component of the received signal in quadrature with the reference signal, Eq. (A.24)
Q_j	component of the sample j of the received signal time series in quadrature with the reference signal
Q_{S_j}	component of the sample j of the received signal time series originating from targets in quadrature with the reference signal
r	radar range
r_d	radar range when used within an integral as the variable of integration
$r_{1,2,3,4}$	radar range of targets 1, 2, 3, and 4, Eq. (2.14) and Fig. 2.14
r_{max}	maximum unambiguous range, Eq. (2.15)
r_v	mixing ratio of water vapor
R	rainfall rate, Eq. (3.7)
R'	gas constant of air
s	unit length
$\mathbf{s}$	unit vector perpendicular to the radar phase fronts
S	fraction of the received signal originating from targets ("signal")

SDR	simultaneous transmit and receive (STAR) differential ratio, Section 6.2.5
S_j	signal fraction of the sample j of the received signal time series
t	time
t_0	time when the transmit pulse was fired, Eq. (2.14) and Fig. 2.14
$t_{1,2,3,4}$	time when the echo from targets 1, 2, 3, and 4 is received, Eq. (2.14) and Fig. 2.14
t_{travel}	time required for the radar pulse to reach the target and come back
T	temperature
T_A	noise temperature of the antenna, Eq. (13.3)
T_{co}	noise temperature of the cosmic microwave background
T_d	dew point temperature
T_v	virtual temperature, Eq. (5.3)
u	east–west horizontal wind component
u_s	east–west wind components at the surface
v	north–south horizontal wind component
v_{DOP}	Doppler velocity of the target, Eqs. (5.2) and (A.8)
v_{max}	Nyquist velocity, Eq. (5.8)
v_s	north–south wind components at the surface
v_{sr}	speed of propagation of the source region
$\mathbf{v}$	three-dimensional wind
$\mathbf{v_{S\text{-}R}}$	storm-relative wind
$\mathbf{v_t}$	three-dimensional velocity of targets
V	integer value representing how a field value at one point is encoded in radar archive files
V_j	sample j of the received signal time series at vertical polarization
w	vertical wind component, or updraft velocity
w_f	reflectivity-weighted average terminal fall speed of hydrometeors with respect to still air
$w_r(D)$	terminal fall speed of a raindrop of diameter D with respect to still air
w_s	vertical air velocity at the surface, Eq. (11.2)
W_a	weighting function with respect to the beam axis describing the angular beam pattern of the radar measurement, Eq. (A.5)
W_r	weighting functions for each cell in range
W_t	weighting functions for each cell in time, Eq. (13.6)
x	east–west distance
X	generic variable or sample
y	north–south distance
Y	generic variable or sample
z	height (everywhere but in Section A.5)
z	generic complex number (Section A.5 only)
$z_{1,2,j}$	generic complex numbers indexed 1, 2, and j
z_{rad}	altitude of radar
z_s	mean sea-level altitude of the surface terrain
Z	radar reflectivity factor, Eq. (3.1)

Z_{dr}	differential reflectivity
Z_e	equivalent radar reflectivity factor, Eq. (3.4)
Z_H	radar reflectivity factor at horizontal polarization
Z_V	radar reflectivity factor at vertical polarization
$\alpha(s)$	absorptivity of a medium at location s along a path, Fig. 2.12
α_1	angle of incidence from the normal to the interface between the two mediums, Fig. 2.9
α_2	angle of exit from the same normal to the interface between the two mediums, Fig. 2.9
β	volume scattering coefficient, Eq. (2.2)
β_D	slope of an exponential drop size distribution
γ	size parameter, Eq. (2.3)
δ_{co}	differential backscattering phase delay, Eq. (6.1)
$\delta\theta$	elevation angle deviation with respect to the beam axis
$\delta\phi$	azimuth angle deviation with respect to the beam axis
$\Delta\varphi$	change in the phase of a target between successive transmit pulses, Eq. (5.7)
$\Delta\phi$	azimuth interval over which pulses are averaged to make a radial
Δr	range interval over which echoes are averaged to make a final range gate
ε	ratio of the gas constants of air and water vapor
ζ	vertical vorticity, everywhere but in Section A.5
ζ	generic complex number, Section A.5 only
$\zeta_{1,2,j}$	generic complex numbers indexed 1, 2, and j
η	radar reflectivity
θ	elevation angle, or angle pointed by the radar with respect to the horizon
θ'	angle with respect to the horizon of the beam as it propagates, Eq. (A.4b)
$\bar{\theta}$	average or center elevation angle
θ_d	elevation angle when used within an integral as the integration variable
θ_j	elevation pointed by the antenna when each transmit pulse j was fired
θ_{beam}	half-power beamwidth of the radar, Eq. (2.11)
θ_{lobes}	half-power width of the sidelobe envelope in elevation
λ	wavelength, Eq. (2.17)
μ	mean of a generic population
ξ_s	scattering efficiency factor, Eq. (2.4)
ρ	air density
$\rho_{X,Y}$	linear correlation coefficient between two time series X and Y, Eqs. (A.14) and (A.27)
$\rho_{X,Y}[l]$	linear correlation coefficient between two time series X and Y at lag l
ρ_{co}, ρ_{HV}	copolar correlation coefficient
ρ_i	density of ice
ρ_s	density of snow (air–ice mixture)
σ	standard deviation of a generic population
σ_b	backscattering cross-section, Eqs. (2.6) and (2.7)
σ_s	standard deviation of a sample, e.g., Eqs. (A.10) and (A.28)
σ_v	spectrum width of the Doppler velocity distribution, Eq. (A.32)

τ	transmit pulse duration
τ_{indep}	time to independence of successive radar measurements, Section 2.4.2
T	transmittance of the atmosphere along the path, Eq. (13.4)
φ	phase of a target, Eq. (5.1)
$\varphi_{1,2,3,4}$	phase of targets 1, 2, 3, and 4, Eq. (2.16) and Fig. 2.14
φ_{HH}	copolar phase of echoes or signal at horizontal polarization
φ_{HV}	cross-polar phase of echoes or signal when the radar transmits at horizontal polarization but receives at vertical polarization
φ_{S_j}	phase of the fraction of the signal originating from targets for sample j
φ_{VH}	cross-polar phase of echoes or signal when the radar transmits at vertical polarization but receives at horizontal polarization
φ_{VV}	copolar phase of echoes or signal at vertical polarization
φ_z	argument of a complex number
ω	wavenumber
ϕ	azimuth angle, or clockwise angle with respect to the north direction pointed by the radar
ϕ'	angle with respect to the local north of the beam as it propagates, Eq. (A.2)
$\bar{\phi}$	average or center azimuth
ϕ_d	azimuth angle when used within an integral as the integration variable
ϕ_j	azimuth pointed by the antenna when each transmit pulse j was fired
ϕ_{lobes}	half-power widths of the sidelobe envelope in azimuth
Φ_{dp}	two-way differential propagation phase
Ψ_0	phase difference at range zero
Ψ_{dp}	differential phase shift between horizontally and vertically polarized returns, Eq. (6.1)

List of acronyms

AMS	American Meteorological Society
AP	anomalous propagation
ARM	Atmospheric Radiation Measurement (facility)
BWER	bounded weak echo region
CALIPSO	Cloud-Aerosol Lidar and Infrared Pathfinder Satellite Observation (satellite)
CAPE	convective available potential energy
CAPPI	constant altitude plan position indicator (radar product)
CEDRIC	Custom Editing and Display of Reduced Information in Cartesian space (software)
CIN	convective inhibition (energy)
DAAC	Distributed Active Archive Center
DART	Data Assimilation Research Testbed

DSD	drop size distribution
EarthCARE	Earth Clouds, Aerosols and Radiation Explorer
EL	equilibrium level
EM	electromagnetic (waves)
EOSDIS	NASA Earth Observing System Data and Information System
GHRC	Global Hydrology Resource Center
GNSS	Global Navigation Satellite System
GPM	Global Precipitation Measurement (satellite mission)
HTI	height-time indicator (radar product)
IET	The Institution of Engineering and Technology
JAXA	Japanese Aerospace Exploration Agency
KFTG	identifier code for the Colorado Front Range radar
KGWX	identifier code for the Columbus Air Force Base radar, Mississippi
KICT	identifier code for the Wichita radar, Kansas
KSRX	identifier code for the Fort Smith radar, Arkansas
LCL	liquid condensation level
LFC	level of free convection
MAPLE	McGill algorithm for prediction by Lagrangian extrapolation
MST	mesosphere–stratosphere–troposphere (radars)
NASA	National Aeronautic and Space Administration (USA)
NCAR	National Center for Atmospheric Research (Boulder, CO, USA)
NOAA	National Oceanic and Atmospheric Administration (USA)
NWS	National Weather Service (USA)
PANT	positive away, negative toward (sign convention for Doppler velocity)
PPI	plan position indicator (radar product)
PRF	pulse repetition frequency
Radar	radio detection and ranging
RASS	radio acoustic sounding system
RHI	range-height indicator
SPC	Storm Prediction Center
TDWR	Terminal Doppler Weather Radar
TRMM	Tropical Rainfall Measuring Mission
TVS	tornado vortex signature
UCAR	University Corporation for Atmospheric Research (Boulder, CO, USA)
UHF	ultra-high frequencies
US	United States
UTC	universal time coordinate
VAD	velocity–azimuth display (radar product)
VIL	vertically integrated liquid (radar product)
WDTB	Weather Decision Training Branch of NOAA
WER	weak echo region
WSR-88D	Model number of the current operational weather surveillance radar used in the United States
Z–R	reflectivity to rainfall (relationship or equation)

Meteorology and radar

If you live in an affluent country, chances are good that one or more radars dedicated to the monitoring of weather take regular measurements of the atmosphere above you. Radar has become a standard instrument in meteorology, joining the thermometer, the radiosonde, and satellite-based imagers as tools used operationally in weather offices. Its images are widely distributed and frequently consulted: in many countries, the web pages showing weather radar images are among the most frequently visited government sites. It is also a key instrument used in research to understand weather phenomena, particularly cloud and precipitation processes. It hence appears that the use of radar in meteorology is here to stay. How and why did this happen?

1.1 How it all started

The year was 1940. World War II raged. The improvement of a decade-old invention, the radar, was being stimulated by the need to detect raiding airplanes and submarines capable of sinking convoys. The radars then transmitted long radio frequency waves and received echoes that bounce off targets, allowing military personnel to detect the enemy at sufficiently long distances to be able to react to the threat. However, at that time, they were huge devices that looked much more like modern-day radio station transmitting antennas, and their angular resolution was poor. A new technological development, the magnetron, provided a solution to this problem by allowing radar to use much shorter wavelengths, microwaves, to achieve the same task; as a result, radar units could become much smaller and be easily moved and installed on aircraft. By the following year, magnetron-based radars were detecting large patches of echoes of unknown origin. It was soon realized that these echoes were caused by precipitation.

War secrecy prevented the publication of such results. Fortunately, during World War II, most meteorological services were part of the military because of the strategic use of weather forecasting. Meteorology personnel were hence shown these images, and immediately realized their potential. Small research groups within the military quickly formed to investigate how this information could be further exploited.

To understand the historical importance of radar as a meteorological tool, it is necessary to remember the state of observing systems at that time. Satellites did not exist, but surface and upper-air observations were taken regularly. This observation network made it possible to map the large- or synoptic-scale patterns of weather systems (>200 km), permitting the detection and the tracking of extratropical cyclones and of anticyclones reasonably well over land, but not as well over oceans. At the other end of the spectrum, human observers at weather stations could

Figure 1.1. Weather radars then and now. Left: picture of a CPS-9 radar, one of the first radars specifically designed for monitoring weather. This particular unit served as the operational radar for the Montreal area from 1954 to 1968 (photo courtesy of Véronique Meunier). Right: the current radar facility in Montreal.

describe the weather at the local scale (<10 km). There was no way to map and study phenomena occurring between these two scales. Radar closed this gap, and the mesoscale was first defined as the scale that could only be studied by this instrument. After the war, radars were put to use in research to observe and understand thunderstorms and their life cycle, and much of what we know of convective storms has been learned using radar observations. Research also focused on understanding cloud and precipitation mechanisms, on what information this new tool could provide, and on how weather radars could be improved. "Radar meteorology" was born.

At the same time, the ability of radar to monitor rapidly developing events such as thunderstorms as well as to track the speed and direction of movement of precipitation systems made it also very interesting for real-time operations. Radars specifically designed to be used for weather monitoring and forecasting were deployed in the early 1950s (Fig. 1.1). The first radar network in the United States was set up in the late 1950s, one of its main roles being to provide advance warning of hurricanes.

1.2 Why radars now

Much has changed since that era. For instance, many more remote sensors have been developed and are being used in meteorology, particularly satellite-based imagers that can obtain data frequently over large portions of the globe. But at the same time, radars used for weather monitoring have undergone considerable transformations (Fig. 1.1), including the ability to obtain wind information using the Doppler effect and to infer the type of echoes observed by transmitting and receiving waves at more than one polarization.

Radar still continues to be used in meteorology nowadays, probably more than ever. Why?

1. Radar remains the best instrument for monitoring the occurrence and movement of precipitation patterns, and people do care about precipitation. A survey about how the public uses and values weather forecasts revealed that precipitation timing, probability, location, type, and intensity forecasts were judged the most important components of weather forecasts along with maximum temperature (Lazo *et al.* 2009). Radars provide the best information on these specifics in the short term (0–6 h), and this alone explains why radar images are so often consulted by the public.
2. Of equal or greater importance, and not mentioned in the survey, is the fact that radar also remains our best tool to detect or infer the presence of many hazardous weather conditions such as severe thunderstorms, hail, and tornadoes. It is able to do so because it is one of the rare instruments that can obtain weather-related information in three dimensions (x, y, z) as a function of time. It can see within storms, a medium through which most electromagnetic radiation cannot penetrate, and can be used to assess their severity using information from the reflectivity, Doppler velocity, and polarization of echoes. Together with satellite imagers, it generally provides us with our last opportunity to recognize whether a previously issued forecast is proving to be wrong and needs to be corrected.
3. Finally, this information is available immediately, and can therefore be used at once.

As a result, there are now networks of radars in many countries whose sole task is weather surveillance, or round-the-clock monitoring of weather events (Fig. 1.2). In fact, at present, the issuance of severe weather warnings is often based on radar observations. Until recently, the limited maximum range of radars (a few hundred kilometers) has hampered their use for larger scale systems. For those situations, geostationary satellite imagers generally provide a much more complete picture, though of clouds rather than of precipitation. But thanks to our increasingly fast communication infrastructure, it is now possible to combine imagery and information from multiple radars in real time and thus obtain useful precipitation monitoring over much larger areas than before. As a result, there is now room for the expansion of the role of radars in meteorology, particularly over large continents, though not as much for island states (Fig. 1.3).

Operational radars also come in a few flavors. In addition to scanning radars that are used for weather surveillance, there are also wind profilers used primarily to derive wind information, and sometimes temperature, above the radar location as a function of height and time. Originally used exclusively for high-atmosphere work, wind profilers have found their place in the arsenal of tools available to meteorologists to supplement radiosonde information. In many countries, they now monitor the current weather and provide data that can be used to initialize numerical prediction models. Finally, there exist many types of research radars that are as varied in size, shape, and mode of operation as are the applications that stimulated their design.

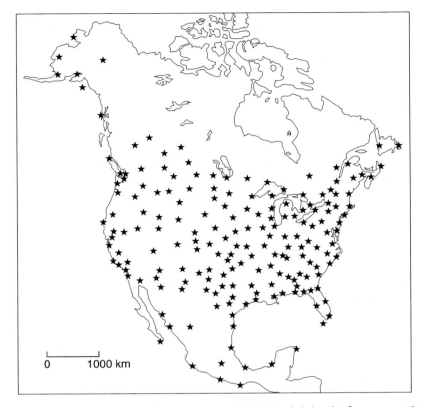

Figure 1.2. North American network of weather surveillance radars circa 2015. Not included in this figure are weather radars dedicated for airport terminal use like the US Terminal Doppler Weather Radars (TDWR) and those owned by nongovernmental organizations such as TV stations.

1.3 Understanding radar observations

The ability of radars to see inside storms and gauge their severity and precipitation intensity makes it a tool of choice for operational meteorologists, researchers, and hydrologists. But radar is also a complex tool, and remote sensing has limitations that must be understood in order to best make sense of the observations. Although considerable progress has been made to quality control the radar data used by meteorologists, radar data processors still cannot succeed in completely removing the many observation artifacts. While detailed technical knowledge of the radar is not required to use its data, a conceptual understanding of the process and limitations of remote sensing measurements is essential to comprehend why things are done the way they are, as well as what problems may arise in the data as a result. This understanding will then help us make the most judicious use of the information.

a) Mesoscale events as seen by a satellite imager

b) Mesoscale events as seen by a single radar

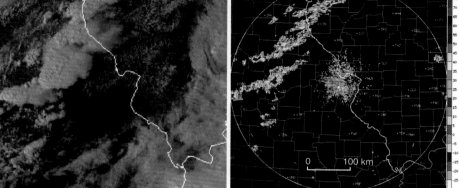

c) Synoptic-scale events as seen by a network of radars

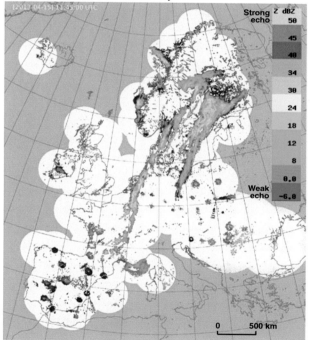

Figure 1.3. Illustration of the information conveyed by radar for mesoscale and synoptic-scale systems. On the top row, the imagery obtained by (a) a satellite-borne visible light imager and (b) a radar is contrasted for mesoscale weather events (© 2006 University Corporation for Atmospheric Research (UCAR), used with permission). On the bottom, a composite image from multiple radars is shown (republished with permission from the American Meteorological Society (AMS) from Huuskonen *et al.* (2014); permission conveyed through Copyright Clearance Center, Inc). For synoptic-scale events, single-radar displays, that generally have diameters of at most 500 km, do not provide a complete picture of weather systems, though this can be achieved by combining multiple radars. At the mesoscale, radar is generally more useful than satellite imagers to determine the location and intensity of convective events.

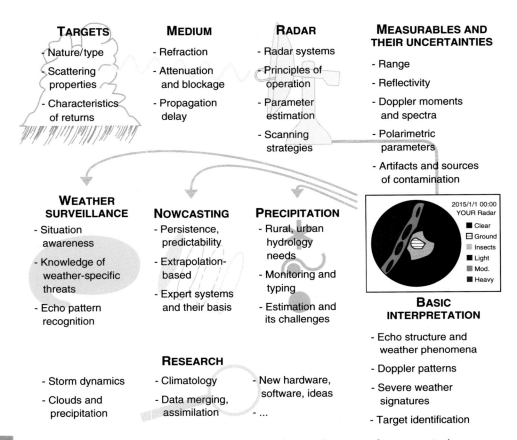

Figure 1.4. Concept map of radar meteorology illustrating the data provided by radars, issues of importance in the interpretation of that data, and the uses of that information in meteorology.

As illustrated in Fig. 1.4, proper understanding of radar data requires some basic knowledge of a variety of topics, including radar operation, propagation, scattering, as well as measurements and their uncertainties. This technical aspect of weather radar measurements has historically been at the center of courses in radar meteorology. But one should keep in mind that the reason we use radars so much is because of the insights they allow us to obtain about weather phenomena. As a result, it is impossible to fully make use of the radar data without understanding both the meteorological phenomena that are observed by this instrument and the value of the information gathered for meteorological or hydrological applications. This is why this book not only presents what radar can observe but also provides some background information on the atmospheric processes shaping the radar imagery and on how we can use the information provided by radar for meteorological and hydrological applications.

The book is structured in three parts. The first section introduces basic radar principles and imagery: what quantities do we measure? How do we recognize different types of targets, or different signatures? On what basis are the inferences made? The second section deals with key applications of radar data, starting with operational uses such as monitoring of convective and widespread systems, precipitation estimation, and short-term forecasting, and progressively shifting to more research-oriented questions including retrievals, cloud and climate uses, and the complicated question of what radar really measures. To provide a support to some of the more technical discussions, an appendix on mathematical and statistical concepts used in radar meteorology completes this book.

1.4 Supplemental readings

For more information on the early years of radar meteorology, the reader is referred to Fleming (1996) that contains many radar-related historical accounts including a chapter on the beginnings of radar meteorology, and other chapters on additional meteorological activities where radar plays a role (e.g., broadcasting). For the most thorough discussion of the history of radar meteorology, Atlas (1990) remains the book of choice: it contains a compendium of knowledge in radar meteorology as of 1990. The early chapters focus on the beginnings of radars at various institutions as well as the historical impacts of radar in various meteorological specializations.

2 Fundamentals of weather radar measurements

2.1 Radar: an active remote sensor

To gather information about the world around us, we rely on a variety of sensing mechanisms. Some, such as touch or taste, are based on *in situ* sensing: the sensor must be in direct contact with an object to gather information about it. Others, such as sight, use remote sensing: the sensor can be some distance away from the object. For remote sensors to work, information must travel between the object and the sensor. Most remote sensors rely on the detection of acoustic waves (sound) or electromagnetic (EM) waves (light, heat, and radio waves, among others) to gather that information.

While the applications of remote sensing are extremely varied, the principles and the process of data gathering are extremely similar. Most remote sensors basically measure the energy received for certain ranges of wavelengths, preferably from a given direction, as a function of time. That energy is either emitted or reflected by the object observed.

Radar, an acronym for RAdio Detection And Ranging, refers to an instrument that emits a strong signal at radio or microwave frequencies and then listens for echoes that occur if the signal reflects off objects known as targets (remember, radar was first a military instrument). And since it provides illumination to the target – like a camera and a flash do, and unlike the way our eyes rely on an external source such as the Sun – it is referred to as an active remote sensor.

Because of the need for an energy source, active remote sensors such as radars tend to be more complex than passive remote sensors such as satellite imagers. But that extra complication comes with benefits. Since we know what was transmitted and when, active remote sensors can make additional measurements compared to passive sensors: How much time elapsed between the transmission of the signal and the reception of the echo? How strong is the signal compared to what was transmitted? Has the frequency or the polarization of the signal changed? These crucial pieces of information give us additional clues on the object being studied, as well as on the medium between the sensor and the object.

From these measurements, and given a model or a mental picture of what we should observe, we interpret the properties measured in order to obtain information on the size, composition, and distance of the object. All remote sensors rely on the combination of detection and interpretation systems. Just as we use our eyes and our brain to understand the scene being observed, artificial sensors use the instrument and the data processing software as a detection–interpretation system. This last step is crucial and poorly recognized: whatever the level of sophistication of the sensor, it is only as good as the assumptions on which the interpretation is based and their implementation in the data processing system. As our eye–brain system can be fooled (optical illusions are proof of that), so can systems based on radar.

What does a radar do?

a. It first generates a strong microwave signal. The task of generating that signal is accomplished by the radar transmitter.
b. It then focuses the signal in one direction, to get information from targets that are located along that specific alone. This is the role of the antenna.
c. It receives the (very) faint echoes from the targets, the intensity of the returned signal being a tiny fraction of what was emitted. The reception of that signal is made possible by the combination of the antenna, which focuses the returns back into the radar system, and the radar receiver.
d. It then extracts as much raw data as possible from the received signals, for example, target range, echo strength, and velocity. Signal processors perform this duty.
e. It processes the raw data to obtain meteorological information. The task of sifting through the large amount of data to produce information of interest to meteorologists is taken care of by the radar product generation hardware and software.
f. It finally displays and disseminates the information, using a radar product display system.

Note that the components involved in the last two functions may or may not be physically located at the radar site; therefore, whether they are considered to be part of the radar system or not is a matter of interpretation. Nevertheless, their existence somewhere is essential to make use of the information obtained. For a quick presentation illustrating what a radar system looks like, consult the electronic supplement e02.1 (figure numbers starting with the letter e refer to electronic material accessible at http://www.cambridge.org/fabry).

After this brief outline of the radar, let us turn our attention to what we intend to use it for: gather information about the atmosphere.

2.2 Microwaves and the atmosphere

We generally are intuitively familiar with many properties of the atmosphere at visible wavelengths. For example, we know that the dry atmosphere is mostly transparent, except for some scattering of blue light that results in the blue sky during the day. We also know that clouds, and to a lesser extent precipitation, scatter light at all wavelengths in the visible part of the spectrum, giving rise to white clouds and thin grayish precipitation trails. However, these properties change considerably with wavelength. The interactions between

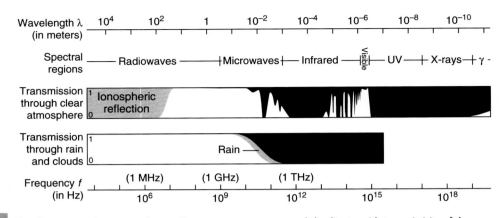

Figure 2.1. The electromagnetic spectrum from radio waves to gamma rays, and the direct zenith transmissivity of the clear atmosphere and of rain and clouds as a function of wavelength. Spectral regions of high transmissivity (all-white areas) are the atmospheric windows.

the atmosphere and microwaves can therefore be very different than those involving radiation at visible wavelengths. Furthermore, both physical and engineering considerations determine what type of information can be obtained by microwave active remote sensing, and they mesh in complicated ways. To understand how remote sensing functions at microwave wavelengths, it then becomes necessary to shed some of our preconceived ideas and try to view the world from the perspective of a radar system.

2.2.1 The radio wave and microwave atmospheric window

For information to travel between an object and a sensor, the medium between the two must allow it. The transparency of the atmosphere to EM waves depends critically on the wavelength of those waves. Figure 2.1 illustrates the direct transmissivity at zenith of the clear and cloudy atmosphere to a variety of EM waves. The atmosphere is transparent only to limited regions of the EM spectrum known as atmospheric windows. These include visible light, narrow regions in the near and thermal infrared, and a large window for the longer microwaves and shorter radio waves (from 1 cm to tens of meters). Radars operate in the latter region, while laser-based systems known as lidars operate in the infrared, visible, and ultraviolet wavelengths. Interestingly, longer microwaves and radio waves will also travel through clouds and most storms without being seriously attenuated. This rare all-weather atmospheric window allows radars to see through storms as well as cover much broader areas than is possible with ground-based optical remote sensing.

The radio wave and microwave atmospheric window covers four orders of magnitude in wavelength, and radars operate over this entire range. In contrast, thermal infrared and visible wavelengths are "only" 1.3 orders of magnitude apart, and this difference is large enough to radically change the atmospheric properties that can be observed and the instruments used to observe them. Based on this finding, one would be right to assume that atmospheric radars working at one end of the radio wave window will be very different and observe very different properties from those working at the other end.

2.2.2 Scattering regimes

Scattering can be loosely defined as the re-radiation of incoming radiation by particles or objects known as scatterers. Partial reflection of EM waves will occur when these propagating waves encounter a medium of a refractive index n_2 different from the medium of refractive index n_1 in which they are traveling. Changes in refractive index in the atmosphere can be sharp or fuzzy. Sharp boundaries occur at the interface between two different bodies. Air–water boundaries between the atmosphere and hydrometeors are examples of sharp boundaries. Fuzzy boundaries occur within the same medium if the properties of that medium change gradually. An example would be air–air boundaries, as air of a given density and moisture may have a slightly different refractive index than an air parcel of different density and moisture nearby.

In the atmosphere, scatterers vary considerably in size, ranging from gas molecules ($\sim10^{-10}$ m) to layers of air at different temperatures and humidity (several meters). Depending on their size and on the wavelength of the incoming EM waves, scatterers will interact differently with this radiation, and bodies that scatter visible light well may not be as effective at microwave wavelengths, and vice versa. Since this aspect is critical to an understanding of what is observed by radar, we investigate it further.

2.2.2.1 Objects with sharp boundaries

Under most circumstances, given the volume of the atmosphere sampled by radar at any instant, multiple targets such as raindrops will be observed simultaneously. In such cases, we are interested in the fraction of the incident radiance scattered per unit length s traveled through the medium. This quantity is known as the volume scattering coefficient β. In the absence of attenuation, the change in flux of energy E at wavelength λ due to scattering is given by

$$\frac{dE(\lambda)}{E(\lambda)} = -\beta ds. \tag{2.1}$$

For spherical scatterers of diameter D in a medium with unit refractive index, the volume scattering coefficient is given by

$$\beta = \frac{\pi}{4} \int_0^\infty N(D) D^2 \xi_s(n(\lambda), D, \lambda) dD, \tag{2.2}$$

where $N(D)\,dD$ is the number of scatterers of diameter D per unit diameter and $\xi_s(n(\lambda),D,\lambda)$ is the scattering efficiency factor, with $n(\lambda)$ being the complex refractive index of the scatterer. Scattering intensity is hence a function of the cross-sectional area of bodies on the path of the radar beam ($\int N(D)D^2\pi/4 dD$), and of how efficiently each of these bodies scatters radiation given its size and refractive index.

The efficiency term is very complex, but there are a few regimes for which it has a relatively simple expression. One of the critical transition points occurs when the circumference of the scatterer πD becomes comparable with the wavelength λ. Let us define the size parameter γ as the ratio of these two terms, that is,

$$\gamma = \frac{\pi D}{\lambda}. \tag{2.3}$$

For scatterers much smaller than the wavelength, ($|n(\lambda)\gamma| < 1$), $\xi_s(n(\lambda),D,\lambda)$ can be approximated by

$$\xi_s(n(\lambda), D, \lambda) = \frac{8}{3}\gamma^4 \left|\frac{n^2(\lambda) - 1}{n^2(\lambda) + 2}\right|^2. \tag{2.4}$$

Scattering occurring in this region is often referred to as Rayleigh scattering or scattering in the Rayleigh region. For particles much smaller than the wavelength, (2.2) therefore becomes

$$\beta = \frac{2\pi^5}{3\lambda^4} \left|\frac{n^2(\lambda) - 1}{n^2(\lambda) + 2}\right|^2 \int_0^\infty N(D)D^6 dD. \tag{2.5}$$

Barring major changes in $n(\lambda)$, it can be seen that under such conditions, scattering is proportional to the sixth power of the size of the object divided by the wavelength to the fourth power. For example, at visible wavelengths, air molecules are much smaller than λ. Hence these molecules scatter shorter wavelengths (blue) much better than longer ones (red), leading to the blue sky. At radar wavelengths (λ often on the order of several centimeters), most atmospheric targets behave as Rayleigh scatterers. Therefore, radars using shorter wavelengths will see increased scattering from precipitation, and this is why it was only when radar wavelengths shortened during World War II that precipitation became visible on radar. It is also one of the reasons why weather radars tend to operate at the shortest wavelength that the atmospheric window allows. Furthermore, with scattering being proportional to D^6, it is clear that larger objects will scatter disproportionately more than smaller ones, and echoes will typically be dominated by the larger objects being illuminated by the radar beam.

For scatterers much larger than the wavelength, $\xi_s(n,D,\lambda)$ loses its direct dependence on wavelength and on diameter and only depends on the refractive index $n(\lambda)$. In this case, and again in the absence of major changes in $n(\lambda)$, it can be seen from (2.2) that scattering is independent of λ and proportional to the area cross-section of each object. At visible wavelengths, this relative independence of β with respect to wavelength results in white clouds. Because this regime describes most of the scattering occurring at visible wavelengths, it is referred to as the optical or the nonselective region.

In between the two regimes, $\xi_s(n,D,\lambda)$ fluctuates wildly as constructive or destructive interference occurs depending on the exact value of the wavelength of the radiation and the size of the object. This region is referred to as the Mie or the resonance region.

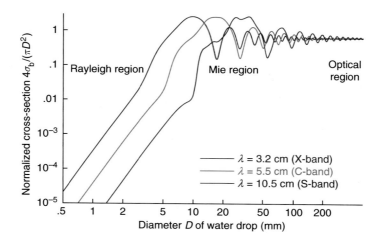

Figure 2.2. Normalized backscattering cross-section of spherical water drops at three radar wavelengths as a function of the drop diameter D. The different scattering regimes can be clearly identified.

Equation (2.5) applies to total scattering. But the directional pattern of scattering is not isotropic, and there can be considerable differences between the amount of energy scattered in one direction and that in another. For radar measurements, one is generally only interested in the scattering coming back to the radar, or backscattering. What hence matters is what is referred to as the backscattering cross-section σ_b of individual targets, expressing the cross-sectional area that a perfect spherical isotropic scatterer would have in the optical scattering regime in order to reflect the same amount of power back to the radar as the target of interest, as well the radar reflectivity η that sums σ_b per unit volume. They can be approximated by:

$$\eta = \int_0^\infty N(\sigma_b)\sigma_b d\sigma_b = \frac{\pi^5}{\lambda^4}\left|\frac{n^2(\lambda)-1}{n^2(\lambda)+2}\right|^2 \int_0^\infty N(D)D^6 dD \tag{2.6}$$

and

$$\eta = \int_0^\infty N(\sigma_b)\sigma_b d\sigma_b = \frac{\pi}{4}\left|\frac{n(\lambda)-1}{n(\lambda)+1}\right|^2 \int_0^\infty N(D)D^2 dD \tag{2.7}$$

for the Rayleigh and optical regimes, respectively. Figure 2.2 illustrates how the backscattering cross-section varies with target size across the Rayleigh, Mie, and optical regimes.

2.2.2.2 Fuzzy refractive index transitions

Reflection from fuzzy refractive index transitions is conceptually more complex to visualize. To properly understand it, one must first remember how mixing in the atmosphere operates in a conceptual manner. Wind and turbulence basically "stretch" and "fold" air

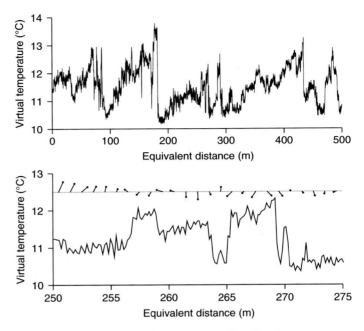

Figure 2.3. Illustration of the fine-scale variability of thermodynamic variables. Top: Virtual temperature measured by an acoustic probe 10 m above the surface as a function of time that has been converted to equivalent distance by multiplying by the wind speed. Bottom: A zoom is made over a 25-m region, and the relative wind perturbations of at most 1.2 m/s are also plotted as vectors above the temperature trace. Clear variability in virtual temperature can be observed at all scales resolved by the probe. The data used in this plot were collected during the Flux Over Snow Surfaces experiment (Mahrt and Vickers 2005) on April 1, 2003, and were provided by Steve Oncley of the National Center for Atmospheric Research.

regions into elements of increasingly smaller scales as turbulence itself cascades from larger whirls to smaller whirls; the final stages of mixing occur at millimeter scales by diffusion as viscosity breaks down turbulence. The net result is that one can observe variability in air properties like temperature and moisture at all scales down to a few millimeters (Fig. 2.3).

The refractive index n of air is not exactly 1, the value in vacuum, as charged particles in atoms interact with EM waves. In the lower atmosphere far from the ionosphere, n at microwave frequencies can be approximated by

$$n = 1 + 10^{-6}\left(\frac{0.776P}{T} + \frac{3.73 \cdot 10^{3}e}{T^{2}}\right), \tag{2.8}$$

where P and e are the air pressure and partial pressure of water vapor, respectively (expressed in Pa), and T is the temperature in degrees Kelvin. Since refractive index depends on temperature and moisture, it also has variability at all scales in the atmosphere. It can be shown that returns from structures of wavelength $\lambda/2$ will interfere in a constructive manner, resulting in a possibly measurable signal even though the variations in refractive index are extremely small. Echoes arising from this process are known as "clear air echoes", as they occur in the absence of clouds or precipitation.

Let us consider the refractive index n at a distance r and $r + L$. When fully developed turbulence shapes the spatial variability of the refractive index and its gradients, it has been shown that the mean-squared refractive index variability $\overline{[n(r) - n(r + L)]^2}$, also called the refractive index structure function, is for distances L between the order of 1 cm and 10 m (Tatarskii 1971):

$$\overline{[n(r) - n(r + L)]^2} = C_n^2 L^{2/3}, \tag{2.9}$$

where C_n^2 is referred to as the refractive index structure parameter. From that parameter, and given the $L^{2/3}$ dependence of the structure function (2.9), the radar reflectivity of the refractive index variability can be determined:

$$\eta \approx 0.38 C_n^2 \lambda^{-1/3}. \tag{2.10}$$

The reflectivity from these clear air echoes is hence weakly dependent on wavelength as opposed to the strong dependence observed for backscattering in the Rayleigh regime (2.6). This difference in behavior explains why radars at shorter wavelengths are better at observing small targets such as precipitation, while longer wavelength radars see clear air echoes more often.

Figure 2.4 puts into perspective what process causes scattering of radiation as a function of the size of the scatterer and the wavelength of the radiation. At traditional radar wavelengths, most targets from air molecules to large hydrometeors are Rayleigh scatterers. Since returns from such targets are proportional to the sixth power of target size, the very small targets that dominate the scattering at visible wavelengths such as air molecules and clouds will be essentially invisible, and hail, raindrops, and snowflakes will be the easiest to detect. And, as wavelength increases, the returns from precipitation targets will diminish rapidly, allowing the relatively weak returns from refractive index gradients to become detectable.

2.2.3 Types of radars

The types and characteristics of radars used for atmospheric remote sensing are dictated by the physics of remote sensing as well as by the applications for which the radar will be used. The key parameter that controls radar capabilities is wavelength: it not only determines what can be observed through its influence on the magnitude of the scattering from different targets but also impacts a variety of technical radar issues. The most important of those other considerations is the width of the radar beam. The angular resolution of a radar depends on the size of the antenna and on the wavelength used. If we consider an antenna made with a parabolic reflector of diameter D_a, a typical antenna design for microwave radars, the angular beamwidth of the radar (in radians) is

$$\theta_{\text{beam}} \approx 1.22 \frac{\lambda}{D_a}. \tag{2.11}$$

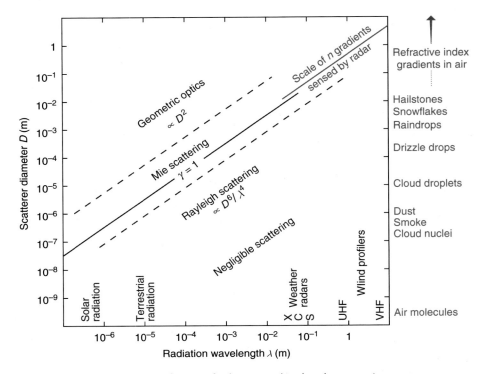

Figure 2.4.   Scattering regimes and processes as a function of radiation wavelength and scatterer size.

This implies that if one chooses a longer wavelength, larger antenna diameters D_a are required for an equivalent angular resolution. For example, (2.11) states that in order to observe the details of storms using a 1° wide beam, a standard angular resolution for weather surveillance radars, an antenna with a diameter of about 70λ is needed. At a 5-cm wavelength, this would correspond to a reasonable size of 3.5 m, while at a 30-cm wavelength, the diameter would have to reach 22.5 m. Hence by necessity, radars with longer wavelengths generally have larger antennas. Larger antennas also help increase sensitivity, and systems with a large antenna and operating at long wavelengths are then capable of observing clear air signals that would have been otherwise undetectable. But smaller wavelength radars can easily have higher resolution and sensitivity to smaller targets. Radars of different wavelengths therefore have different strengths and weaknesses, making them suitable for different applications.

For reasons mentioned above, it is generally impractical to have narrow-beam radars at large wavelengths ($\lambda > 20$ cm), especially if the antenna is expected to move. However, relatively weak echoes from refractive index gradients can be observed in the troposphere and above, making it possible to obtain information under all weather conditions. In particular, it is possible to measure wind speed and direction using the Doppler effect.

Figure 2.5.   The Middle and Upper Atmosphere (MU) radar at Shigaraki, Japan, is a large MST wind profiler made of an array of antennas.

To take advantage of this possibility, a special class of radars that uses a large antenna array on the ground to measure winds over the sensor as a function of height has been designed (Fig. 2.5). These radars are referred to as wind profilers. Wind profilers come in a few varieties determined by their wavelength: shorter wavelength profilers ($\lambda \approx 30$ cm) are limited to observations in the boundary layer unless precipitation is present, while increasingly longer wavelength profilers (from 75 cm to 6 m and beyond) can make useful observations at increasing heights until they can become "mesosphere–stratosphere–troposphere" (MST) radars (Hocking 2011). Some profilers are also accompanied by loudspeakers that generate sound with an acoustic wavelength near $\lambda/2$. The profiler can then detect those sound waves and measure their speed. Since the speed of sound is a function of (virtual) temperature, temperature profiles can be derived from that information. The latter is known as a radio acoustic sounding system, or RASS.

As we decrease the wavelength to several centimeters, a large antenna of several meters in diameter can yield a good angular resolution. Yet these radars are generally not severely affected by attenuation because we are still well within the microwave atmospheric window, and data can be obtained up to a few hundred kilometers. In parallel, refractive index gradient echoes have become extremely weak, leaving room for the largest Rayleigh targets, precipitation-size hydrometeors (Fig. 2.4). As a result, this class of radars is generally used for weather surveillance, and most national weather radar networks

Examples of weather surveillance radars. Clockwise from upper left: The Denver (USA) area WSR-88D (KFTG) from outside; the antenna and pedestal of KFTG from inside the radome; a television-owned radar on top of the Rockefeller Center in New York City; a scanning radar system in the State of São Paulo, Brazil; and the S-Pol transportable research radar in Colorado.

use either 5- or 10-cm wavelength radars (Fig. 2.6). The choice between these two wavelengths is generally based on financial considerations, past experience with a given system, climatic conditions, and the perceived importance of the issue of attenuation by strong precipitation. As we go to somewhat smaller wavelengths (e.g., 3 cm), we can find mobile radars used for research applications: designed to be small enough to be moved on roads, they sacrifice long-range coverage because of heavier attenuation for the ability to get close to storms of interest and obtain high-resolution imagery as a result (Fig. 2.7, left; supplement e02.2).

If we keep decreasing the wavelength to 1 cm and below, penetration through precipitation becomes too limited for weather surveillance applications. And because precipitation-size targets like snowflakes and raindrops become Mie scatterers, their echo becomes less strong and more difficult to interpret, allowing smaller-sized objects such as cloud droplets

Figure 2.7. Left: Mobile research radars waiting for storms in Oklahoma. Right: A short wavelength cloud radar deployed on a boat (cloud radar photo courtesy of Prosensing Inc.).

and ice crystals to become the main target of interest. In parallel, these radars become very sensitive systems that can be very compact (Fig. 2.7, right). They are hence generally used at short ranges or for special applications where very high sensitivity (for cloud detection), very high resolution (for research), or compact radar systems (spaceborne radars) are required.

2.2.4 Naming convention of radar bands

The microwave spectrum is subdivided into bands. During World War II, these bands were given "letter names," partly in order to complicate possible spying attempts of the radar programs. Specialists trained to understand this naming convention kept using it after the war, and its usage has survived to this day, at least in the civilian radar community. Figure 2.8 illustrates the wavelengths associated with these radar bands as well as some of the other uses of the microwave and radio EM spectrum. Because of frequency allocation regulations, weather radars are generally limited to about 10 narrow bands, each roughly a factor of two apart in wavelength from its nearest neighbor. This implies that most weather surveillance radars have to be either at S-band ($\lambda \approx 10.5$ cm), such as the US WSR-88D, or at C-band ($\lambda \approx 5.5$ cm), as adopted in many European countries. Occasionally, shorter range or lower-cost radars operate at X-band ($\lambda \approx 3.2$ cm). Shorter wavelength bands are primarily used by research or space-borne radars.

The great variety of radars is therefore dictated by the measurement physics at microwave frequencies, the nature and type of targets to be observed, and the intended application. Wind profilers and weather radars have distinct but complementary roles, the former measuring winds with height from the echo of refractive index gradients and the

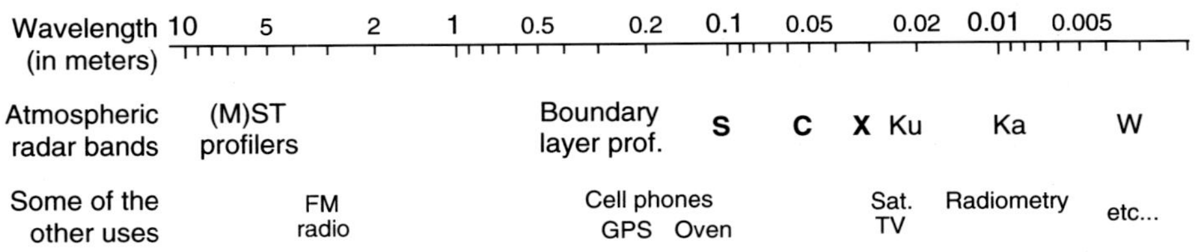

Wavelength (in meters)	10	5		2	1	0.5		0.2	0.1	0.05		0.02	0.01	0.005	
Atmospheric radar bands	(M)ST profilers					Boundary layer prof.			**S**	**C**	**X**	Ku		Ka	W
Some of the other uses		FM radio						Cell phones GPS Oven				Sat. TV	Radiometry		etc...

Figure 2.8. Naming convention and wavelength of radar bands used in meteorology. The most frequently used bands are indicated in bold.

latter generally tracking precipitation-size targets associated with storms. In addition to these operationally used systems, there are many specialized radars filling niches, all functioning based on the same principles.

2.3 Propagation in the atmosphere

The atmosphere not only provides the targets to the radar but also forms the medium in which radar waves travel. As these waves propagate through the atmosphere, they are both refracted and attenuated.

2.3.1 Refraction

Radar waves are refracted, or change direction, when the refractive index n of the medium in which they travel, air, changes. Most introductory physics textbooks tell us that n in the atmosphere is essentially 1, the value in vacuum. In reality, as seen in (2.8), n depends on pressure, temperature, and humidity. Near sea level, n is approximately 1.0003, or 300 ppm over 1. While n is indeed close to 1, the gradual change of the refractive index of air between the surface and deep space will significantly bend the trajectory of radar waves.

Consider the following example. If the refractive index of a medium changes abruptly between n_1 and n_2, Snell's law tells us that $n_1 \sin(\alpha_1) = n_2 \sin(\alpha_2)$, where α_1 is the angle of incidence from the normal to the interface between the two mediums and α_2 is the angle of exit from the same normal to the interface (Fig. 2.9). If the computation of $\sin(\alpha_2)$ leads to a number greater than 1, total reflection occurs instead of refraction. For a radar scanning at an elevation angle of 0.5°, one can easily calculate that a sharp n gradient, parallel to the surface, of only 40 ppm is sufficient for complete reflection to occur: this is because $\sin(\alpha_1) = \sin(89.5°)$ equals 0.999962, a number already very close to 1.

Under most circumstances though, n changes gradually with height. Let us consider the refractivity $N = 10^6(n-1)$ that corresponds to the quantity in parentheses in (2.8). On average, in the free troposphere, N decreases exponentially with height, that

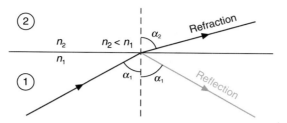

Figure 2.9. Geometry of reflection (orange) and refraction (red) by a long refractive index boundary between two media (1) and (2) given radiation coming from (1) in the bottom left.

is, $N = N_o \exp(-z/H_o)$, where N_o is the refractivity at the surface and H_o is the scale height (or the e-folding distance) of the atmosphere for refractivity, and ranges between 6 and 9 km. Closer to the surface, dN/dz varies considerably, especially at night and during convection, when low-level temperature and humidity profiles are generally the most variable. Given a value for dn/dz, one can show (Bean and Dutton 1966) that the radar beam is bent toward the region of higher n with a radius of curvature r such that

$$\frac{1}{r} = -\frac{1}{n}\frac{dn}{dz}\cos\theta', \qquad (2.12)$$

where θ' is the angle between the direction of propagation and the horizontal (Fig. 2.10, left). This curvature is hence maximized for near-horizontal beams. For typical values of dn/dz, the radar beam will generally curve slightly back toward the surface, but with a larger radius of curvature than the radius of the Earth.

Calculations of the altitude of the beam as a function of range could be challenging because we are trying to measure the distance between a curving beam and a curving Earth surface. To simplify these calculations, we can "straighten" the trajectory of the beam by assuming that the problem at hand is equivalent to one where the radar wave would be propagating in a straight line over a planet with an equivalent radius $k_e a_e$, where a_e is the radius of the Earth, that is,

$$\underbrace{\frac{1}{a_e}}_{\substack{\text{Earth's}\\\text{curvature}}} - \underbrace{\left(-\frac{1}{n}\frac{dn}{dz}\cos\theta'\right)}_{\text{Beam curvature}} = \underbrace{\frac{1}{k_e a_e}}_{\substack{\text{Modified Earth's}\\\text{curvature}}} - \underbrace{0}_{\substack{\text{Straight}\\\text{beam}}}. \qquad (2.13)$$

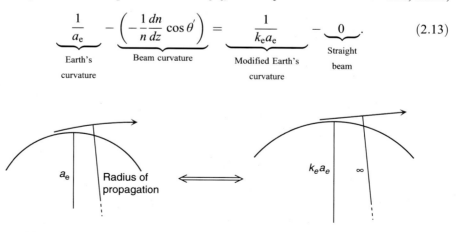

Figure 2.10. Illustration of the enlarged Earth radius approximation for simplifying the calculation of radar beam trajectories.

In the first kilometer above the surface, dn/dz is on average around $-40 \times 10^{-6} \text{ km}^{-1}$. If this value is used in (2.13) for low elevation angles ($\cos \theta' \approx 1$), we get $k_e \approx 4/3$. This "4/3 Earth radius approximation" is commonly used to compute the propagation of radio waves near the surface, for example for microwave links. In radar meteorology, this can be used to compute where the radar beam will intercept the surface under normal propagation conditions and hence where ground echoes should be observed (see Appendix A.1 for details). Higher in the atmosphere, on average, dn/dz decreases with height, and so would the beam curvature. As a result, the 4/3 Earth radius approximation tends to slightly underestimate the beam height, though the difference is a small fraction of a beam width.

When dn/dz has a value close to normal (dn/dz around $-40 \times 10^{-6} \text{ km}^{-1}$, or dN/dz around -40 km^{-1}), we have normal propagation conditions. Anomalous propagation occurs when dn/dz departs significantly from this normal value, such as when $dn/dz > 0$ or $dn/dz < -80 \times 10^{-6} \text{ km}^{-1}$. Sometimes, the rate of decrease of n is smaller than average. In such cases, we have sub-refraction conditions, meaning that the radar beam curvature is reduced, resulting in a beam height that is higher than expected. Sub-refraction conditions occur with a faster than normal decrease of temperature with height, and/or a slower than normal decrease (or an increase) of moisture with height. These conditions are difficult to obtain meteorologically, and sub-refraction is rarely strong. Its effects are thus relatively benign.

At other times, the rate of decrease of n becomes larger than average. When this occurs, super-refraction is observed as the radar beam curvature is enhanced and the beam is lower than expected. Super-refraction occurs when we have a slower than normal decrease (or an increase) of temperature with height, and/or a faster than normal decrease of moisture with height. Nighttime temperature inversions, fronts, storm outflows, and warm air flowing over cold water will all cause super-refraction of radar and radio waves. A special case of super-refraction occurs when n decreases so rapidly that dn/dz becomes more negative than minus the curvature of the Earth ($-1/a_e$), or $-157 \times 10^{-6} \text{ km}^{-1}$. When this happens, the radius of curvature of the beam becomes smaller than that of the Earth's surface, and the beam bends back toward the ground. If this layer of highly negative dn/dz is sufficiently thick and/or low in the atmosphere, "trapping" of radar waves occurs and a significant fraction of the radar beam cannot penetrate through that layer.

One of the consequences of super-refraction conditions near the surface is that a greater than normal fraction of the radar beam will hit the ground. More ground targets than usual appear on the radar display (supplement e02.3), and these unusual ground targets are referred to as anomalous propagation echoes, often abbreviated as anoprop, or simply AP (Fig. 2.11).

2.3.2 Attenuation

In addition to being refracted, microwaves can be attenuated. Attenuation is the main factor that dictates where the microwave atmospheric window ends at shorter wavelengths

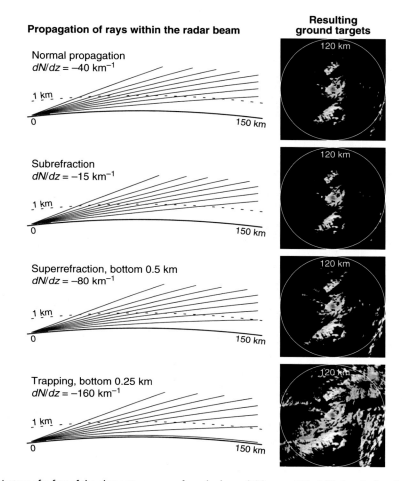

Propagation of rays within the radar beam

Resulting ground targets

Normal propagation
$dN/dz = -40$ km^{-1}

1 km

0   150 km

Subrefraction
$dN/dz = -15$ km^{-1}

1 km

0   150 km

Superrefraction, bottom 0.5 km
$dN/dz = -80$ km^{-1}

1 km

0   150 km

Trapping, bottom 0.25 km
$dN/dz = -160$ km^{-1}

1 km

0   150 km

Figure 2.11. Left: Trajectory of a few of the elements, or rays, of a radar beam (1° beam width, 0.5° elevation) under different propagation conditions (normal, sub-refraction, super-refraction near the surface, and trapping near the surface). Right: Resulting simulated ground echo pattern over a 240-km by 240-km area centered on the radar, in this case for the McGill radar in Montreal, Canada.

(Fig. 2.1). There are many sources causing the attenuation of microwaves by the atmosphere (Fig. 2.12). Oxygen and water vapor absorb some microwaves, particularly at certain bands (1.3 cm for H_2O, 5 mm for O_2). Attenuation caused by liquid cloud and precipitation increases steadily with frequency. At frequencies below 3 GHz (wavelengths greater than 10 cm), attenuation is small and easily correctable under all but the most extreme circumstances. As frequency increases, it becomes an increasing concern. For example, at 3-cm wavelength, echoes behind a thunderstorm cell 5–10 km wide will appear approximately 95% weaker than they would have been if the cell had not been present (Figs. 2.12 and 2.13). Shorter wavelengths will suffer from attenuation even more, increasingly limiting both their maximum range and their use for

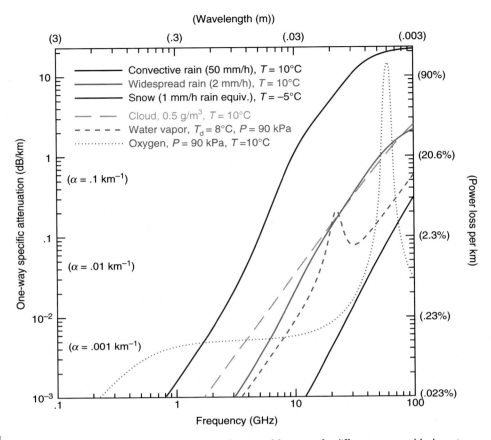

 Figure 2.12.   Specific one-way attenuation or absorptivity a as a function of frequency for different gases and hydrometeors.

precipitation monitoring. A last source of attenuation is a wet radome (the golf-ball-shaped dome that usually covers the antenna, e.g., Fig. 2.6). When a film of water covers the radome, large attenuation may result, again especially at shorter wavelengths (supplement e02.4). Hence, attenuation can be a severe limitation when attempting to use radar data quantitatively. Supplement e02.5 provides more quantitative information on attenuation.

2.4 Basic radar measurements

We will now take a first look at how to interpret the information contained in the returned signal.

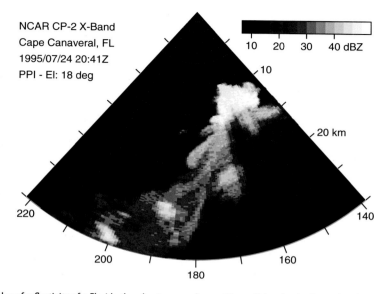

NCAR CP-2 X-Band
Cape Canaveral, FL
1995/07/24 20:41Z
PPI - El: 18 deg

Figure 2.13. Sector display of reflectivity of a Florida thunderstorm as observed by an X-band radar located at the upper tip of the image (brighter colors = stronger echoes). The main cell with a 5-km diameter is sufficiently intense so as to cause an obvious reduction of the echoes behind it.

2.4.1 Timing, range, and radial velocity

At this point, we focus on how radars can measure the strength and radial velocity of individual targets located at different ranges from the radar. Figure 2.14 concentrates all the information in one complex illustration. The best way to fully comprehend how this functions is by alternating between reading a few sentences of the following text and examining the relevant portion of Figure 2.14.

When the radar transmits a pulse, the antenna is pointing in a particular direction in azimuth and elevation. Even if the antenna moves, its pointing direction can be considered constant over the short time taken by a radar pulse to reach its targets and return to the receiver. In this example, the radar beam in Figure 2.14a illuminates four point targets of different sizes at ranges from r_1 to r_4 from the radar. These targets are moving in different directions, two getting closer to the radar and one going further away, while the other is stationary.

Now let us consider how the radar pulse propagates in range as a function of time (Fig. 2.14b). Starting at time t_0, the radar fires a high-power pulse of microwaves for a short duration τ. This pulse, illustrated by the area striped vertically, propagates in the atmosphere at the speed of light in air c/n. When the pulse reaches the four targets, a small portion of its energy will be reflected back and will reach the radar sometime after

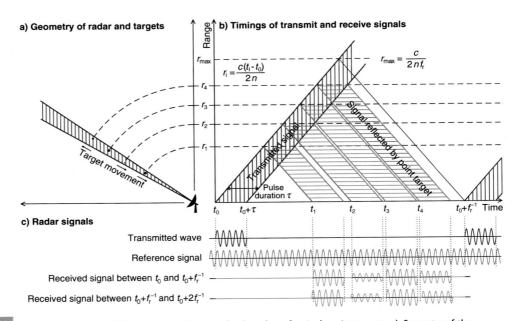

Figure 2.14. Pictorial description of the nature and timings of radar echoes for single point targets. a) Geometry of the problem with the radar at the origin, the beam axis striped vertically, and four point targets, the two nearest approaching from the radar and the furthest receding from it. b) Time-range plot of the transmitted pulse of duration τ (striped vertically) and the reflected signals from each point target (striped horizontally). c) Illustration of the transmitted wave, the reference signal with respect to which the phase of targets is computed, and the received signals as a function of time (or range) from two successive radar pulses. From Fabry and Keeler (2003), © Copyright 2003 AMS.

t_0. The echoes from each target are illustrated by the area striped horizontally. The times of arrival t_1 to t_4 of each echo can then be used to infer the ranges r_1 to r_4 of each target using

$$r_i = \frac{c(t_i - t_0)}{2n}.$$ (2.14)

After some time f_{r}^{-1}, a new pulse is fired. The frequency f_{r} is referred to as the pulse repetition frequency. It also determines the maximum unambiguous range up to which a radar can observe targets, because echoes arriving from beyond that range are indistinguishable from those originating from closer ranges but caused by the next pulse. That unambiguous range is

$$r_{\max} = \frac{c}{2nf_{\mathrm{r}}}.$$ (2.15)

Finally, let us focus on the signals as perceived by the radar (Fig. 2.14c). As mentioned earlier, the radar periodically emits a short burst of microwaves ("transmitted wave" line on Fig. 2.14c). The echo from each target consists of a carbon copy of the transmitted wave, whose time of arrival is determined by the range of the target. The amplitude of the received signal from each target depends both on the size and properties of the target and on the target's position with respect to the center of the beam. If we compare the phase difference φ_i between the return and a reference signal in phase with the transmitted wave, we obtain

$$\varphi_i = 2\pi f(t_0 - t_i) = -\frac{4\pi f n}{c} r_i. \tag{2.16}$$

This phase difference φ_i is known as the phase of the radar return. If the transmit frequency f and the refractive index of the atmosphere n are constant, changes in the phase of the signal from one transmit pulse to the next are directly caused by changes in the range of the target. This forms the basis of the radar's ability to measure the radial velocity of targets. The radar transmit frequency also determines the wavelength λ of the radar pulse in the atmosphere via

$$\lambda = \frac{c}{nf}. \tag{2.17}$$

2.4.2 Radar returns from a distributed target

In precipitation, one can find many billions of targets within the sampling volume illuminated by the radar at any given instant. The echoes from each of these targets combine to make the signal received by the radar.

Figure 2.15 and the supplement e02.6 illustrate that process. In Figure 2.15, one can see three diagrams, the first one showing the transmit wave and the other two illustrating the return signal toward the radar at two different times, t_1 and t_2. Transmitted single-frequency radar pulses (1) illuminate a volume of space (the rectangular area in the two bottom illustrations). Each target in that volume, represented by discs of various sizes, will scatter a portion of the wave back to the receiver, the amplitude and phase of the return depending on the size and the position of the targets (2). The echoes of all the targets will interfere and combine with one another (3), resulting in a single wave returning back to the receiver. That wave has two main properties, an amplitude A (4) and a phase φ (5) with respect to the transmit wave. The radar return can also be expressed as a complex number $(A \exp(i\varphi)) = (I + iQ)$, where i is the square root of -1, and illustrated by a vector (5). If targets move in time (note the shifts in their position between t_2 and t_1), the phase of their individual returns will change. In parallel, the returns from stronger targets have larger amplitudes.

It is from these measurements of amplitude and phase of targets that the basic measurements of weather radars, reflectivity and Doppler velocity, are derived. The amplitude of the signal is related to the square root of the return power P_r ($P_r = A^2/2$), and from P_r both the backscattering cross-section of targets (2.6) and the radar reflectivity can be derived. The phase φ measured is the result of the combination of the returns from each target and has little meaning for weather targets; its rate of change, however, depends on how rapidly

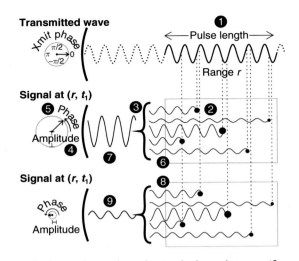

Figure 2.15. Illustration of the processes leading to the weather radar signals observed at a specific range. Top: The wave transmitted by the radar illuminates a volume of space from which echoes are received. Middle: Each target in the volume scatters an echo that combines with the others to make up the signal observed by the radar at time t_1. Bottom: Same as in the middle, but at a time t_2 shortly after t_1. (See the text and the supplement e02.6 for additional details.) From Fabry and Keeler (2003), © Copyright 2003 AMS.

the combination of targets observed in the sampling volume approaches or recedes from the radar.

Because the signal received from all the targets is in the form of waves (Fig. 2.15), it will vary depending on how all the returns from individual targets combine together. If the returns of many targets interfere constructively (6), the resulting signal will be strong (7); if they interfere destructively (8), the resulting signal will be weak (9). The radar signal amplitude hence fluctuates in time as targets move with respect to each other in and out of spatial configurations where constructive and destructive interference dominates in the signal received by the radar. The rate at which this echo fluctuation occurs depends on the speed with which targets within the sampling volume reshuffle with respect to each other, and not on their absolute velocity.

Because of echo fluctuation, any single measurement of reflectivity may bear little resemblance to the mean reflectivity that would be observed by averaging the signals from all possible combinations of target positions (Marshall and Hitschfeld 1953). But as more measurements are taken, the echo amplitude will evolve in time as targets reshuffle in position within the sampling volume. The time to independence is defined as the time required for targets to sufficiently move with respect to each other that a new measurement of the target reflectivity will be statistically independent from the previous one. This time to independence will increase with wavelength λ and decrease with the width of the velocity distribution σ_v according to $\tau_{\text{indep}} = \lambda/(4\pi^{1/2}\sigma_v)$. For microwave weather surveillance radars and normal shear and turbulence within the pulse volume, it is of the order of 10 ms at S-band (10.5-cm wavelength).

To obtain a good estimate of mean reflectivity, it is necessary to get many independent estimates because of the fluctuating nature of radar echoes from precipitation. Presently, this fact is the main constraint that determines the amount of time that radars will point in the same direction: one must wait long enough to obtain a sufficient number of independent measurements in order to achieve the needed accuracy in reflectivity. Because of these echo fluctuations, the radar must observe in a particular direction for a certain amount of time (tens to hundreds of milliseconds) before one can obtain a good reflectivity measurement.

2.5 Weather surveillance

Radar waves can travel long distances in the atmosphere, and radars provide information almost instantaneously. This is why they play an important role in weather surveillance as well as in meteorological research. Atmospheric phenomena are typically three-dimensional, although their vertical extent is generally much smaller than their horizontal one. This fact plus the spherical shape of the Earth impose a limit on how far ground-based radars can observe weather (Fig. 2.16). In addition, information obtained at a single elevation and azimuth is clearly insufficient to appropriately gauge the type and severity of atmospheric phenomena. Proper weather surveillance hence requires that the radar use scanning strategies that sample the atmosphere as thoroughly as possible. But weather phenomena can move and evolve rapidly. The scanning strategy used by radar is hence always the result of a trade-off between data accuracy, coverage, and temporal resolution.

In general, scanning strategies are composed of a set of three basic scans. The first one is when the antenna scans in azimuth at a fixed elevation (Fig. 2.17, left). Raw images generated by this scan are called plan position indicators, or PPIs. By extension, the term PPI is also used to refer to the scan itself and not only to its display. PPIs are commonly used when horizontal resolution is privileged over vertical resolution. The second type of scan is when the antenna scans in elevation at a fixed azimuth (Fig. 2.17, center). Raw images generated by this scan are called range-height indicators, or RHIs. RHIs are used primarily in research systems when improved vertical resolution is needed. The last type

Figure 2.16. Vertical section to scale of the geometry of radar measurements. A 1° wide beam at an elevation of 0.5° illuminates the troposphere. By the time the beam reaches a range of 250 km, it samples a large volume at the top of the troposphere; beyond that range, the radar rapidly becomes blind to the weather underneath.

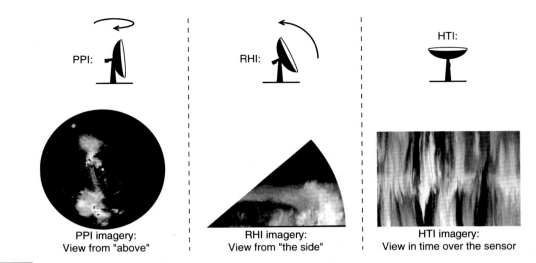

Figure 2.17.   Illustration of three basic radar scans: PPI (left), RHI (middle), and HTI (right).

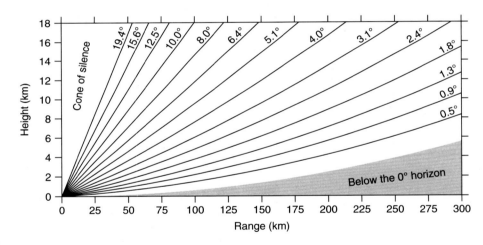

Figure 2.18.   Height of the beam axis as a function of range for the 14 elevation angles used in the Volume Coverage Pattern #12 scanned by the US WSR-88D radars.

of "scan" is when the antenna elevation and azimuth are kept fixed for a certain amount of time. These range-time indicators are primarily used by radars with limited scanning capabilities such as wind profilers or to obtain data at very high resolution in range and/or time. When the elevation is the zenith or nadir, these are also called height-time indicators (HTIs, Fig. 2.17, right).

Weather surveillance radars typically use sets of PPIs to scan the atmosphere. Figure 2.18 shows one of the scanning strategies used by the US WSR-88Ds in stormy conditions. There exists no standard approach for determining how to best select the elevation angles to be scanned. There are, however, some basic principles that are used. For example, scans made at low elevation angles cover the largest areas and therefore contain the most information. Furthermore, if the effect of the curvature of the Earth is neglected, it can be shown that to have the same number of measurements in height at far range as at near range, angles must be spaced such that $\tan(\theta)$ increases geometrically. Finally, there is a practical limit to the number of angles that weather surveillance radars can scan: because severe weather may evolve rapidly, the whole measurement cycle should not exceed 5–10 min. As a result, the number of angles will rarely exceed 25, and the maximum elevation is generally below 40°. Weather surveillance radars are hence often blind to weather immediately above them, and this unmeasured volume is known as the cone of silence.

As we saw in this chapter, what radar can observe is dictated by an intricate blend of scattering and propagation physics, instrument specifications, and scanning strategy. To learn more about radars and how they work, you will have to wait until Chapter 13. For now, we have presented the minimum required physical and technical considerations of radars to understand the information obtained from them. We can now turn our attention toward data processing and interpretation considerations.

3   Radar reflectivity and products

Echo signals received by the radar in its natural spherical coordinate during its scanning cycle must be converted into products that can be used meteorologically. A product can be described as a field of a quantity of meteorological interest set on a specific projection. Both the field and the projection aspects of products are discussed in this chapter. Historically, products were designed to be used by forecasters, but more and more, radar-derived information is being used by other algorithms such as numerical models in conjunction with data from other sources.

Maps or images of many fields can be generated using radar data. In this chapter, we focus on those based on the intensity of the returned signal. To be of use, that intensity must be converted into a meteorologically meaningful quantity, like rainfall rate. It is achieved using the radar equation.

3.1 The radar equation

The average amount of power received from a given range $P_r(r)$ can be interpreted quantitatively after making some assumptions about target properties. In Chapter 2, we saw that most precipitation targets are much smaller than the usual radar wavelengths and therefore behave as Rayleigh scatterers. For such targets, it is convenient to define a quantity Z called the radar reflectivity factor per unit volume such that

$$Z = \int_0^\infty N(D)D^6 dD, \qquad (3.1)$$

$N(D)$ being the number of hydrometeors of diameter D per unit volume. This radar reflectivity factor is one of the target properties we seek to measure with meteorological radars. For historical reasons, and to avoid having measurements with large negative exponents all the time, Z is expressed in nonstandard units of mm^6/m^3. One 1-mm drop per cubic meter has a reflectivity of 1 mm^6/m^3 while three 2-mm drops per cubic meter have a combined reflectivity of 3×2^6 or 192 mm^6/m^3.

Information about the radar reflectivity factor can be obtained from measurements by using the radar equation. There exist many variations of the radar equation that differ somewhat in the assumptions that are used. A convenient version for radars with parabolic antennas is

$$P_r = \frac{1.22^2 0.55^2 10^{-18} \pi^7 c}{1024 \log_e(2)} \underbrace{\frac{P_t \tau D_a^2}{\lambda^4}}_{\substack{\text{Radar} \\ \text{parameters}}} \underbrace{\frac{T(0,r)^2}{r^2}}_{\text{Path}} \underbrace{|K|^2 Z}_{\substack{\text{Target} \\ \text{properties}}}, \tag{3.2}$$

where P_t is the power of the transmit pulse and τ its duration, D_a is the diameter of the antenna, T is the transmittance of the atmosphere along the path between the radar at range 0 and the sampling volume at range r, and $|K|^2$ is the dielectric constant of the scatterers and is related to the complex index of refraction of the hydrometeor $n(\lambda)$ via

$$|K|^2 = \left| \frac{n(\lambda)^2 - 1}{n(\lambda)^2 + 2} \right|^2. \tag{3.3}$$

All quantities in (3.2) are in SI units except Z which is in mm^6/m^3. If one is interested, the derivation of (3.2) is explained in the electronic supplement e03.1. Equation (3.2) is valid if all particles have the same dielectric constant, are spherical, behave as Rayleigh scatterers, and are randomly distributed with a mean density that does not vary within the sampling volume. As indicated, the equation has four terms: constants, a term depending on radar parameters (transmit power, pulse length, antenna size, and wavelength), a term depending on the properties of the path to the targets (partial blockage, attenuation, and range), and a final term depending on target properties (dielectric constant and reflectivity factor). Given radar parameters and good knowledge about path effects, one can hence obtain quantitative information about the radar reflectivity factor Z from the strength of the echo received.

Equation (3.2) implies that the weakest echo that can be detected by radars is a function of range. For example, the same received power will be received from a target of reflectivity Z_1 at range r_1 as from a target of reflectivity $Z_2 = 4Z_1$ at range $r_2 = 2r_1$. For this reason, the minimum detectable signal of a radar is generally quoted as a certain reflectivity at a particular reference range.

3.2 The equivalent reflectivity factor

Equation (3.2) can be used to obtain Z provided one knows the dielectric constant $|K^2|$ of targets. This is not always the case: one does not always know for certain what the target is made of; it could even be made of a mixture of several substances. Table 3.1 lists the dielectric constants of some of the meteorological targets observed by radar. Between liquid water and solid ice, the dielectric constant varies by a factor of 5. Therefore, the power received by the radar for a liquid water target will be five times stronger than for a solid ice target with a similar reflectivity factor. The situation is even more complicated with snow as it is a mixture of ice and air, and what constitutes the edge of a snowflake is a matter of debate. From here on, we choose to define the edge of the snowflake as the smallest convex spheroid that includes all the ice in the snowflake. With typical densities

Table 3.1. Dielectric constants at radar wavelengths for different targets

Target type	Dielectric constant $\|K\|^2$
Liquid water (cloud, drizzle, rain)	0.93
Solid ice ($\rho_i = 920$ kg/m^3)	0.176
Air–ice mixture (snow) of density ρ_s, $\rho_s < 200$ kg/m^3	$\approx 0.205 \, (\rho_s/\rho_i)^2$

between 50 and 100 kg/m^3, snowflakes contain a lot of air and have smaller dielectric constants than pure ice; but being a lot larger, they also have much larger reflectivity than solid ice particles of the same mass (recall that the radar reflectivity factor defined in (3.1) is proportional to D^6). When these two effects are combined, we find that snowflakes reflect slightly more radar energy than solid ice particles of the same mass.

Because of uncertainties about the nature of the target, or whether it is a Rayleigh scatterer or not, we define a new quantity called the equivalent reflectivity factor Z_e such that

$$P_r = \underbrace{\frac{1.22^2 0.55^2 10^{-18} \pi^7 c}{1024 \log_e(2)}}_{\text{Constants}} \underbrace{\frac{P_t \tau D_a^2}{\lambda^4}}_{\substack{\text{Radar} \\ \text{parameters}}} \underbrace{\frac{T(0,r)^2}{r^2}}_{\text{Path}} \underbrace{\|K_w\|^2 Z_e}_{\substack{\text{Target} \\ \text{properties}}}, \tag{3.4}$$

with $\|K_w\|^2$ being the dielectric constant of liquid water (0.93). If attenuation is negligible or corrected for, it is this equivalent reflectivity factor Z_e that is measured by the radar. Note that $Z_e = Z$ if the targets are made of liquid water and are small compared with the radar wavelength.

The (equivalent) reflectivity factor of targets can span many orders of magnitude. For example, a liquid cloud has a typical reflectivity factor of 0.01 mm^6/m^3, while the hail core of a thunderstorm can have reflectivity factors exceeding 10^6 mm^6/m^3. For convenience, we generally express reflectivity factors in units of decibels (dB, or tenths of orders of magnitude) of mm^6/m^3 such that

$$dBZ = 10 \log_{10}(Z). \tag{3.5}$$

With this convention, the thunderstorm would have a reflectivity factor of $10 \log_{10}(10^6) = 60$ dBZ while the reflectivity factor of the liquid cloud would be $10 \log_{10}(0.01) = -20$ dBZ. Note that a negative dBZ value does not mean a negative reflectivity factor, but simply a reflectivity factor of less than 1 mm^6/m^3 (see Box 3.1). On most radar displays, the quantity referred to simply as "reflectivity" is the equivalent reflectivity factor deduced by the radar using (3.2) and (3.4), with the intensity scale having dBZ units (Fig. 3.1). A good way to rapidly become comfortable with the meaning of the dBZ values shown on radar displays is to use the reference values provided in Table 3.2.

| Box 3.1 | **Units of Reflectivity in Radar Meteorology** |

Units and symbols used in the radar meteorology literature can be confusing. First the radar reflectivity factor Z and similar quantities have unusual units of mm^6/mm^3. Then, we do not generally express the reflectivity in linear Z (or mm^6/mm^3) units, but instead in "dBZ," or $10\log_{10}(Z/(1\ mm^6/m^3))$. Complications will grow when we later introduce dual-polarization variables. One of these new variables is "differential reflectivity" Z_{dr}; it is only a difference of two reflectivity measurements when one uses logarithmic units, but is strictly speaking a ratio of reflectivity in linear units. But then we nevertheless display and express it in dB. The possibility for confusion becomes very real when one starts using these symbols in equations: should we use the linear or the dB value? Some authors are trying to resolve the ambiguity by giving lowercase symbols (e.g., z) to linear values, and uppercase symbols (e.g., Z) to dB values. It is being resisted by meteorologists and traditionalists (both of which I am) because we are familiar with Z as a linear quantity in equations, and because z is already commonly used for height in meteorology, introducing additional confusion. Hence, from here onward, I will use Z and other reflectivity-based variables in equations in linear units. Be aware that others may not use a similar convention.

In parallel, one should always be careful whether reflectivity values should be used and manipulated in their linear or logarithmic forms. See Appendix A.2 for a more detailed discussion on this issue.

3.3 Reflectivity factor and rain rate

Consider the following two equations for the reflectivity factor Z and the rainfall rate R:

$$Z = \int_0^\infty N(D)D^6 dD \tag{3.6}$$

and

$$R = \frac{\pi}{6} \int_0^\infty N(D)D^3 w_r(D)dD. \tag{3.7}$$

In these two equations, $N(D)$ is the number of raindrops per unit volume, also referred to as the raindrop size distribution, and $w_r(D)$ is the fall speed of a raindrop of diameter D. The drop size distribution varies with rainfall rate and depends on the processes influencing precipitation growth. Hence, there exists no mathematical function linking Z and R. For example, it can be easily shown that one 2-mm diameter drop falling at 7° m/s has a reflectivity factor similar to 64 drops of 1-mm diameter drops falling at 4 m/s, and yet the

Table 3.2. Typical values of equivalent radar reflectivity factors from different targets

Target type	Z_e (dBZ)
Light drizzle; insects	0
Moderate drizzle; a few raindrops; light snow; migrating birds	10
Light rain or moderate snow, typical of widespread precipitation (1 mm/h)	25
Moderate rain, strong for widespread precipitation (5 mm/h)	35
Heavy rain from a convective shower (20 mm/h)	45
Hail or very heavy rain, peak of thunderstorms (100+ mm/h)	55
Moderate to severe hail	>60

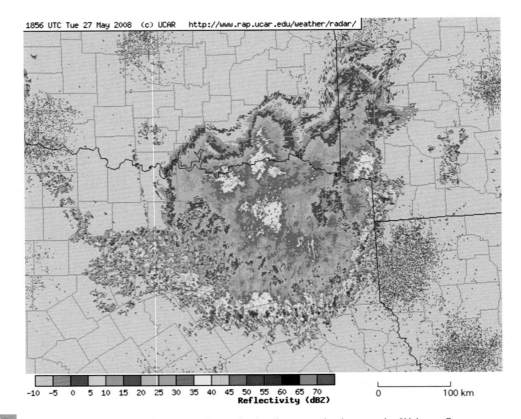

Figure 3.1. Reflectivity composite image made by combining the data from several radars near the Oklahoma–Texas–Arkansas border in the United States. A "smiley convective complex" can be observed. © 2008 UCAR, used with permission.

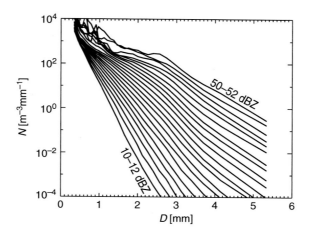

Figure 3.2. Average drop size distribution for all 2-dB reflectivity factor intervals between 10–12 and 50–52 dBZ computed from 5 years of observations in Montreal, Canada. Republished with permission of the AMS from Lee and Zawadzki (2005b); permission conveyed through Copyright Clearance Center, Inc.

two rainfall rates are very different. Because of the D^6 weighting on Z, reflectivity factor is dictated by the larger targets and is relatively unaffected by the smaller drops, even though those contribute a large fraction of the rainfall.

However, on average, drop size distributions vary systematically with reflectivity (Fig. 3.2) and precipitation intensity. This fact has allowed us to derive Z–R relationships that can be used to convert reflectivity factors into rainfall rates. For example, Marshall and Palmer (1948) fitted exponential functions through their drop size distributions and obtained

$$N(D) = 8000\, e^{(-4.1R^{-0.21}D)}, \tag{3.8}$$

where $N(D)$ is in m^{-3}/mm, R is in mm/h, and D is in mm. Using data on the fall speed of drops as a function of diameter (e.g., Gunn and Kinzer 1949), one gets

$$Z = 300R^{1.5}, \tag{3.9}$$

a relationship Joss and Waldvogel (1970) also found from their disdrometer data. There exist many Z–R relationships, most of them power laws of the form $Z = aR^b$ as in (3.9) but with different a and b coefficients. These change depending on the dynamical and microphysical processes controlling precipitation formation, and hence vary a little from one region to another. For example, precipitation generated by the warm rain process such as drizzle or shallow showers results in smaller drops; hence, areas that receive more warm rain, such as coastal or tropical regions, will have climatological Z–R relationships with smaller a coefficients (a weaker reflectivity will be observed for the same rain rate because of the smaller drops). Examples of other Z–R relationships include the influential Marshall–Palmer Z–R relationship

$Z = 200R^{1.6}$ originally derived in the early 1950s from (3.8) but adjusted to better fit the observations available at the time, and the WSR-88D default relationship $Z = 300R^{1.4}$ whose lower exponent makes it more suitable in regions where a greater fraction of the precipitation is from deep convection. In Chapter 9, we will revisit Z–R relationships and drop size distributions in the context of trying to properly estimate rainfall by radar.

3.4 Radar products

Data collected by radar must be displayed if people are to interpret them. Weather surveillance radars collect data in three spatial dimensions and in time, and those must be rendered on 2-D displays. Some data manipulations or transformations are therefore required. In addition, data such as reflectivity may be further processed in order to extract more information. The results of these manipulations are known as radar products.

As seen in Chapter 2, the most common scanning strategy of weather surveillance radars is to perform several scans in azimuth or PPIs, each at a different elevation angle. The simplest product is hence a PPI display of reflectivity or Doppler velocity at a specific elevation angle. One of the disadvantages of a PPI display is that there is a systematic change in the observation height with range: at close range, one observes echoes near the surface, while at far range echoes originate from much higher in the atmosphere. A way to compensate for this is to build a constant altitude PPI, or CAPPI (Fig. 3.3). A CAPPI is

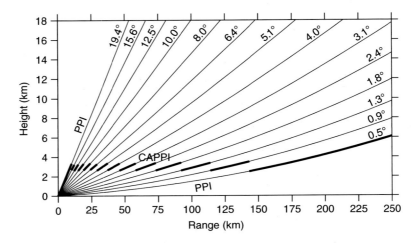

Figure 3.3. Illustration of the generation of a simple 3-km CAPPI. Thin lines represent the PPIs available while the thick lines highlight the elevations from which data are used to make the CAPPI. More sophisticated CAPPIs can also be made by interpolating data from two elevation angles for each pixel.

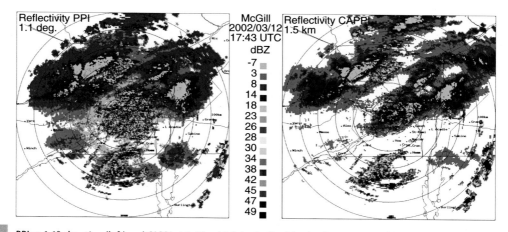

Figure 3.4. PPI at 1.1° elevation (left) and CAPPI at 1.5 km (right) of reflectivity for the same snowfall event with no ground echo filtering applied. On both images, textured intense echoes are due to ground targets while the more uniform weaker echoes are from snow. Note that the PPI on the left has (1) stronger ground targets at close range, (2) stronger snow echoes at close range, and (3) weaker snow echoes at far range than the CAPPI on the right, all resulting from the changing altitude of observation with range. Range rings are every 20 km.

made by selecting some information from all elevation angles in an effort to generate a display representing the weather at a specific level above ground or sea level. Besides removing the change in observation height with range, CAPPIs naturally show fewer ground echoes (Fig. 3.4). They do not solve the problem of the widening beam with range and can sometimes lead to misinterpretations if they are centered at a height near the 0°C isotherm where the transformation of snow to rain leads to unusually strong echoes. Use of CAPPIs is much more common in some countries than in others, partly for historical reasons, and partly because CAPPIs require a larger number of elevation angles (at least a dozen) in order to maintain a small deviation from the desired height, and because in some countries data are collected only at a few elevation angles.

One of the great strengths of radar is its ability to make measurements in three dimensions. Several products are hence designed to extract that information. Two of the most useful are as follows:

1. Vertical cross sections (Fig. 3.5): Using data from all the elevation angles, vertical cross sections can be made along user-selected paths. They are very useful in providing us with a glimpse of the height structure of a storm or a set of storms. How high does the storm core extend? Is it stretching vertically or is it slanted? Are there cores aloft or voids at low levels? How high is the melting layer? Those are all questions that one can answer by making vertical cross sections or RHI scans.

2. Vertically integrated liquid (VIL, Fig. 3.6): The VIL product attempts to estimate the amount of precipitation already condensed in a vertical column using the reflectivity data. VIL is expressed in kg/m^2 or in mm of water (1 mm of water

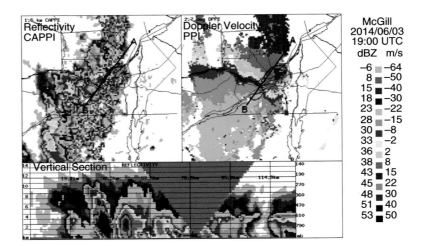

Figure 3.5. Example of a vertical cross-section product below with contextual information above. On the reflectivity CAPPI in the upper left, and the Doppler velocity PPI in the upper right, one can perceive near its center the path from A (northeast) to B (southwest) over which the cross section is made. Below the two horizontal maps is the actual vertical cross section of reflectivity, the point A being on the left and B on the right of the image. Two vertical scales are provided, one in height (in km) on the left and one in pressure (in hPa) on the right. This vertical cross section done over a user-selected 125 km line shows storm cells on the left extending up to 12 km altitude, with lighter precipitation on the right.

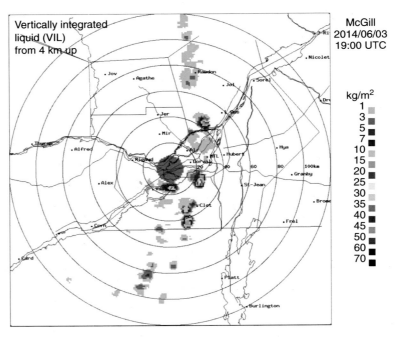

Figure 3.6. A VIL product for the same time as Fig. 3.5. The amount of condensed water estimated from 4 km to the top of the atmosphere is shown in this particular example. Cells to the east appear to have in that layer up to 30 mm of precipitation-size hydrometeors ready to fall.

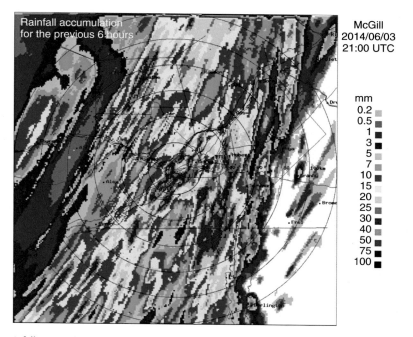

Rainfall accumulation
for the previous 6 hours

McGill
2014/06/03
21:00 UTC

mm
0.2
0.5
1
3
5
7
10
15
20
25
30
40
50
75
100

Figure 3.7. The 6-h rainfall accumulation computed over a 240-km × 240-km area for the event used in Figs. 3.5 and 3.6. Note how for such a convective event, within the same forecast area, some regions have received little or no precipitation (e.g., on the right of the image) while others have received more than 50 mm (e.g., just southwest of the radar), an amount exceeding half a month of normal precipitation in this area.

spread over 1 m^2 weighs 1 kg). This product is used qualitatively to quickly and effectively identify the strong precipitation cells (Clark and Greene 1972). It is also used semiquantitatively in a variety of other severe weather products aimed at estimating hail size or wind gust potential.

Scanning radar data are also collected at regular time intervals of a few minutes. This information in time can also be used in a variety of ways. Past information may be simply displayed either via animations (supplement e03.2) or by synthesizing information on a single image such as past storm tracks. Additional processing may also be performed on past images, for example, to generate precipitation accumulations over the past 1–24 h (Fig. 3.7). Finally, the current and past radar information on the location of a storm can be used to issue very short-term forecasts of a variety of quantities: storm tracks, rainfall accumulations or probabilities, etc. These are just some examples of products that are possible using radar data, and imagination is the only limit. We will see how those products are used in later chapters.

It should be stressed that products are not strictly 2-D maps of a given measured or derived quantity. Many algorithms look through radar data in search of specific signatures worth identifying and warning for, especially in severe weather. Algorithms sifting through 3-D radar data include hail detection algorithms, the detection of weak echo regions within convective storms or of slanted strong echoes, and cell identification and tracking, not to mention many data quality assurance algorithms that attempt to clean radar data from contamination of various origins.

4   Reflectivity patterns

The basic physics and measurement strategy of weather radar having been established, it is time to turn our attention on the radar observations themselves, as their interpretation forms the core of radar meteorology. Radar echoes have a variety of origins and shapes, each associated with, as well as caused by, the processes that generate the targets being observed. Understanding and differentiating echo patterns must hence come with an understanding of the phenomena that give rise to them.

4.1 Types of targets

Targets observed by radar can be separated into three broad categories:

1. Precipitating weather targets: These are what naturally come to mind when one thinks of weather radar observations: rain, drizzle, snow, hail, etc.
2. Nonprecipitating weather targets: These include ice clouds, water clouds, and refractive index gradients causing clear-air echoes. The ability to detect each of these will depend on radar sensitivity and wavelength.
3. Nonmeteorological targets: These include a large variety of targets. Some of these are important to detect such as ash; some others such as insects end up providing information that will prove to be useful for meteorological purposes. Unwanted targets for most meteorological purposes include birds, airplanes, and the ground and sea surfaces. Echoes from unwanted targets are referred to as clutter. Note that someone's clutter may be someone else's signal: aviation radar users refer to weather echoes as clutter!

Despite the large variety of targets, most of them can be recognized by carefully examining the horizontal and vertical structure of their echo. To achieve this, the processes governing their formation and evolution must be well understood. We must therefore explore in some detail the microphysics and dynamics of precipitation formation.

4.2 Precipitation processes: a quick overview

In a nutshell, dynamical processes determine the structure and spatial extent of the areas of vertical motion required to generate clouds and precipitation, while microphysical processes control the way water vapor supersaturation, resulting from upward motion, is ultimately transferred into precipitation.

There are several instabilities and processes in the atmosphere that may cause enough vertical motion to generate precipitation. The two most important are the baroclinic instability and the convective instability. Baroclinic instability arises from the existence of a meridional temperature gradient in an atmosphere in quasigeostrophic equilibrium and possessing static stability (AMS 2014). It leads to the formation of large-scale (a few thousand kilometers) midlatitude cyclones. Except near cold fronts, vertical velocities are at most a few tens of centimeters per second, much slower than the fall speed of precipitation. Because of the large extent and weak intensity of updrafts, precipitation is light and covers broad areas. The variability in the vertical velocities, and hence in the precipitation, is primarily in height; as a result, precipitation generally has a stratiform (layered) structure, though it is by no means horizontally uniform. At the other end of the spectrum, convective instability occurs when a lighter fluid such as warm air overcomes the influence of viscous forces and rises over a heavier fluid such as colder air. Especially when condensation occurs, air that is convectively unstable can ascend (and descend) at speeds of tens of meters per second. Because of the quickness with which convection occurs, it is often a localized phenomenon. Convective instability gives rise to showers and thundershowers, and individual cells have a horizontal scale of the order of 10 km. The fast and localized upward motion gives to convective rain its high intensity and complex 3-D structure. Other dynamical instabilities and forcing mechanisms such as topography can affect the formation and growth of precipitation, but baroclinic and convective instabilities are both the most common and the most extreme in characteristics.

Echoes will look very different depending on the dynamical process dominating precipitation formation. If the strength of the vertical velocity is small and its spatial variability is limited as is the case when baroclinic instability dominates the precipitation process, echoes will have large horizontal extent, weaker horizontal variability, and weaker intensities (Fig. 4.1, left). On the other hand, if vertical velocities are large and have large horizontal variability, echoes will be more intense and have cellular structures (Fig. 4.1, right) organized in a variety of patterns depending on the forcing mechanism and the horizontal and vertical structures of temperature and winds. Finally, convective weather evolves much more rapidly in time than baroclinic weather. The typical lifetime of a convective cell is less than an hour; this is one of the main reasons why we want radars to make new volume scans every few minutes.

Precipitation formation is shaped not only by the dynamics of updrafts but also by the microphysical processes controlling the growth of precipitation-size particles or hydrometeors. Liquid hydrometeors can form by two processes. In the warm rain process, vapor condenses on condensation nuclei to form cloud droplets; these cloud droplets then coalesce to form drizzle drops or raindrops. This process is generally dominant when cloud top temperatures are warmer than $-10°C$. In the cold rain or Bergeron–Findeisen process, vapor deposits on ice crystals that then grow and aggregate with each other to form snowflakes; these snowflakes fall and then melt when the temperature exceeds $0°C$, forming raindrops. Note that these two processes often coexist in the same volume when the updraft is large enough (e.g., Politovitch and Bernstein 1995, Zawadzki *et al.* 2000), and

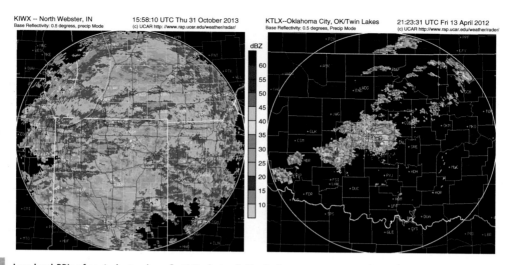

Figure 4.1. Low-level PPIs of equivalent radar reflectivity factor (left) of a low-pressure system driven by baroclinic instability and (right) of thunderstorm cells generated by convective instability. The outer ring is 230 km from the radar. © 2012, 2013 UCAR, used with permission.

it is common for snowflakes to collide with and capture liquid cloud droplets. While dynamical processes tend to have a greater impact on the horizontal structure of targets, microphysical processes mostly affect their vertical structure.

4.3 The vertical structure of rain echoes

Precipitation dynamics affect the spatial structure and intensity of precipitation, and hence of precipitation echoes. Microphysics dictates the type of hydrometeors observed, the processes affecting their growth, as well as their growth rate; this shapes the vertical structure of reflectivity. As a result, both precipitation dynamics and microphysics determine the structure of weather echoes. Specifically, four main types of vertical structures for weather echoes are identified (Fig. 4.2), each associated with a particular combination of precipitation process.

1. Stratiform warm rain: The warm rain process in the absence of significant updrafts leads to the formation of weak and shallow precipitation made of unusually small drops given the rainfall rate achieved. This is what we refer to as drizzle. In general, the reflectivity of drizzle increases steadily as precipitation falls down, and it reaches at most 10–20 dBZ. When drizzle forms in subzero temperatures, it is supercooled and can pose a threat to aviation.

2. Stratiform cold rain: Large-scale precipitation generated by the cold rain process is the most common form of precipitation outside of equatorial and tropical regions. It often

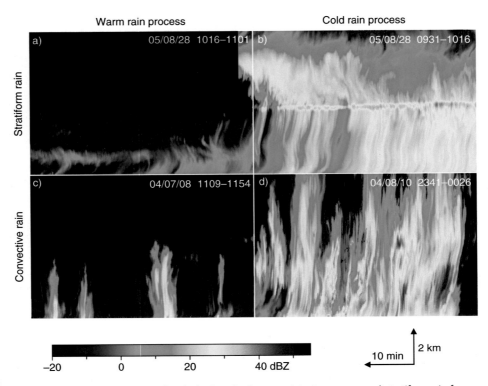

Figure 4.2. Vertical structure of echoes associated with the four dominant precipitation processes: a) stratiform rain from the warm rain process; b) stratiform rain from the cold rain process; c) convective rain from the warm rain process; d) convective rain dominated by the cold rain process and riming. The examples of stratiform (a) warm rain process and (b) cold rain process are part of the same event.

extends throughout most of the troposphere. Its vertical profile of reflectivity is characterized by a steady increase of Z in snow as height decreases ($dZ/dz < 0$), followed by a local maximum (the "bright band") and fairly constant Z in rain. The reflectivity in rain rarely exceeds 40 dBZ (10 mm/h), as stronger rainfall requires strong updrafts that are not compatible with the release of baroclinic or symmetric instabilities. *The bright band and other characteristic structures of these echoes will be covered in more details in Section 4.4.*

3. Convective warm rain: This type of echo is typically associated with showers, either from isolated cells or as part of a line along a cold front. The echoes are not very tall as showers generally do not extend beyond the $-10°C$ level. Nevertheless, precipitation can be very strong and reflectivities may exceed 50 dBZ (50 mm/h).

4. Convective "cold" rain: Convective precipitation where the cold rain process plays a significant role is generally associated with thunderstorms. Nonetheless, cloud and precipitation growth by the warm rain process is also important in updrafts. Just as for convective warm rain echoes, thunderstorm echoes are either isolated or form lines or clusters. They are often followed by stratiform rain (Fig. 4.3). Echoes are tall, intense, and generally have a cellular or globular appearance. Reflectivities may exceed 55 dBZ in rain. If they exceed 60 dBZ, hail is likely to be present.

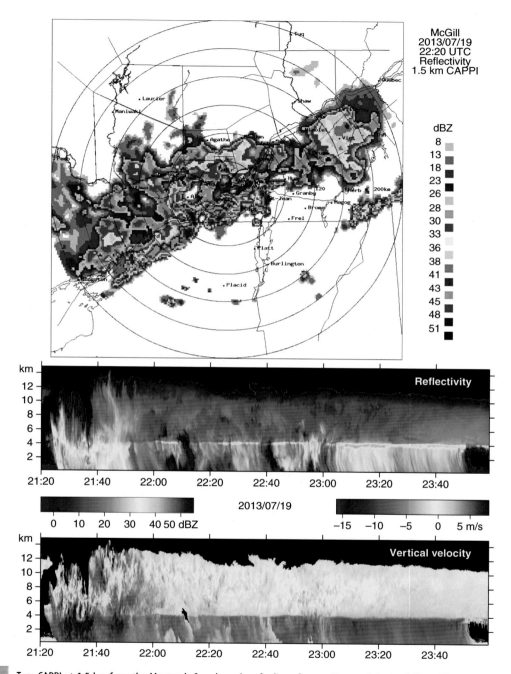

Figure 4.3. Top: CAPPI at 1.5 km from the Montreal, Canada, radar of a line of convective precipitation followed by increasingly stratiform rain. Bottom: Time–height sections of reflectivity and vertical velocity taken at the radar site for the same storm. In the vertical velocity image, note the bubbly texture in convection (>40 dBZ echoes on the CAPPI above, corresponding to echoes prior to about 21:50 UTC on the time–height section below) and the sharp transition in fall velocity between snow and rain in the stratiform region.

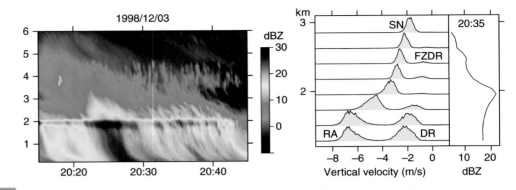

Figure 4.4. Left: Time–height section of reflectivity during what looks like a cold rain event. Right: Distributions of the vertical velocity of targets (positive upward) at eight different heights at 20:35. Two precipitation modes can be observed: one made of snow (SN) falling at 2 m/s and melting to rain (RA, 7 m/s), and one made of supercooled drizzle (FZDR) remaining liquid (drizzle, DR) as it goes through the 0°C level. The presence of growing drizzle drops also implies the existence of undetected liquid clouds. On the far right, the corresponding vertical profile of reflectivity is plotted.

While echoes fall predominantly in one of those categories, one should remember that several processes may coexist at any one time: stratiform precipitation may include small cells of convection. Furthermore, over some temperature ranges, warm rain, liquid clouds, and frozen precipitation may coexist and interact (Fig. 4.4). Despite the temptation, one must not take the four categories above as absolutes, but rather as end points of a distribution of possibilities. That being said, the probability of observing echoes matching one of those end points is greater than that of observing something in between.

4.4 Radar signatures of widespread rain

Figure 4.5 illustrates the vertical and horizontal structure of precipitation echoes for a large-scale rain event. It is a very rich image, and we examine several of its characteristics in more detail.

4.4.1 Precipitation trails and their movement

In large midlatitude systems, precipitation is often generated by the cold rain, or Bergeron–Findeisen, process: first, ice crystals nucleate; they then grow by diffusion, accretion, and aggregation; and finally, they melt into raindrops that then collide with cloud droplets and other raindrops. Near the maximum height of echoes – the echo top – some regions will experience stronger updrafts than others. More water vapor is available for diffusion on

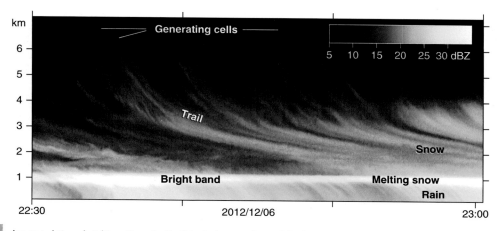

Figure 4.5. Annotated time–height section of reflectivity in large-scale precipitation.

growing crystals, and this releases heat, further enhancing the strengths of updrafts and the growth of crystals. This positive feedback leads to what are known as generating cells, from which more intense echoes seem to originate. Once snowflakes are formed, they fall and form precipitation trails. These may be more or less visible depending on the intensity and the horizontal scale of the generating cells. Trails will often persist until the precipitation reaches the surface, though they tend to fan horizontally because of the spread in the fall speeds of the hydrometeors that compose them.

Individual raindrops and snowflakes follow a trajectory dictated by their fall speed and the strength and direction of the wind advecting them. As a result, localized irregularities in reflectivity will follow the wind horizontally and move down with the fall speed of hydrometeors. But at any given instant, the trail of hydrometeors originating from a given source region, whether a generating cell or another phenomenon, forms a shape dictated by the difference between the wind $\mathbf{v}$ at a given level, the speed of propagation of the source region $\mathbf{v}_{sr}$, and the average fall speed of hydrometeors (Fig. 4.6). If echoes are tracked in time horizontally and not vertically, it is the trails of precipitation that will be followed, not the precipitation irregularities that are moving down and following the wind. It is hence a mistake to believe that one can deduce the winds by following the horizontal movement of precipitation echoes. The displacement velocity of a precipitation trail is independent of height and is dictated by the speed of the source region. The source region moves at a speed corresponding to the wind at that level plus its propagation speed with respect to air. One can hence measure the movement of weather echoes by observing them at any height level as long as there are targets present: the resulting velocity will be the same. This fact will prove convenient to facilitate making short-term forecasts on the basis of their observed movement. Also note that in Fig. 4.5, the trails near echo top seem to follow a roller coaster-like trajectory. This unusual occurrence is due to the presence of very strong localized waves that are forcing snowflakes up and down.

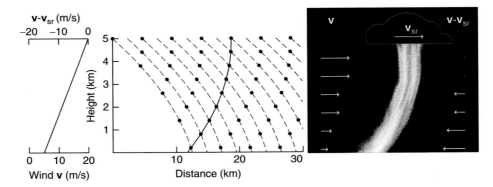

Figure 4.6. Illustration of the formation of precipitation trails. Precipitation falling at 5 m/s is released continuously from a point source (center image). Dashed lines show the trajectory followed by some of the drops as they are pushed by the winds shown on the left. The position of these drops every 4 min is also shown by dots. At any instant, all the drops released from the source region form a trail (solid line) which moves at the velocity of the source region while the drops constituting the trail move horizontally at the speed of the wind at their level. Right: Real trail measured by a vertically pointing radar subjected to the wind profile **v** (brighter colors corresponding to stronger echoes). As on the illustration to the left, the deviation of the slope of the trail from the vertical at a given level is due to the velocity difference between the source region and the wind at that level. Adapted from Marshall (1953) and Fabry (1993), © Copyright 1993 AMS.

4.4.2 The bright band signature of melting snow

In the middle of Fig. 4.5, one can spot a band of enhanced reflectivity. This "bright band" is the reflectivity signature of melting snow (Fig. 4.7) and is particularly visible in large-scale rain events. It is hence useful to locate the level of the 0°C isotherm in precipitation. However, if radar measurements at the bright band level are interpreted as a measure of surface rainfall, the amount of precipitation occurring will be grossly overestimated which will result in flood alarms being incorrectly issued.

Bright bands are caused by the combination of three factors (Fig. 4.8): (1) a change in the dielectric constant $|K|^2$ of melting hydrometeors from the small values for snow (see Table 3.1) to the much larger value for liquid water; (2) a change in the size of hydrometeors from large fluffy snowflakes to small dense raindrops; and (3) an associated change in the fall speed of these hydrometeors resulting in a reduction in the concentration of hydrometeors per unit volume. When melting starts, snowflakes become wet. This does not appreciably change their size or their density, but it does increase their dielectric constant. As a result, the strength of the echo from melting snow increases. This continues until snowflakes are almost totally melted, at which point their increasingly fragile melting ice structure collapses into raindrops with some remnants of ice. The size of the melting snow rapidly diminishes, as a result of which its density increases, leading to a rapid acceleration of the hydrometeor. Since precipitation flux (mass concentration

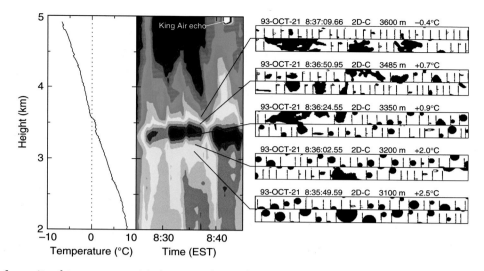

Figure 4.7. Composite of temperature and hydrometeor shape taken during an ascent through the melting layer by the University of Wyoming King Air instrumented aircraft and simultaneous reflectivity observations of that melting layer by a UHF wind profiler. The temperature profile taken during the ascent through precipitation is shown on the left. In the center, the radar reflectivity is presented in height–time coordinates with warmer colors representing stronger echoes. The radar echo of the aircraft 5 min after crossing the bright band (8:42 EST) can also be seen. On the right, portions of the 2D-C cloud probe records taken at five heights (in snow, 1/5th of the bright band, bright band peak, 4/5th of the bright band, and in rain) are presented. In these records 0.8-mm wide, raindrops appear as disks, snowflakes as complex rough shapes, while melting snowflakes are somewhere in between depending on the degree of melting. From Fabry *et al.* (1995), © Copyright 1995 AMS.

times velocity) is more or less conserved through melting, an increase in fall velocity implies a reduction in the mass concentration (or number) of particles per unit volume. The reduction in both size and concentration results in a decrease of the echo strength at the final stages of melting. As a result, melting snow has an equivalent reflectivity factor greater than both the snow above and the rain below, resulting in a bright band signature. Condensation of vapor on melting snowflakes, aggregation of melting snowflakes, and breakup of melted hydrometeors also occur, but they generally play minor roles in the shape and intensity of the bright band signature.

If snowflake densities are much higher than usual (dotted line, Fig. 4.8), the bright band becomes less pronounced. This is the case when frozen hydrometeors grow primarily by accretion (the collection of supercooled cloud droplets) or riming, a common situation if there is significant convection at subzero temperatures. As a result, the bright band signature becomes progressively weaker as the precipitation becomes more convective (Fig. 4.3), and it becomes essentially invisible in deep convection.

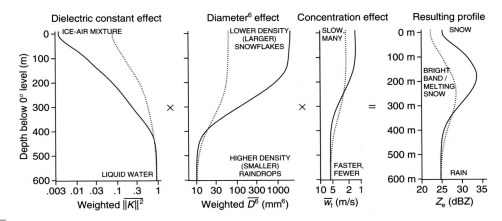

Figure 4.8. Computed height profiles of the contribution of the three key factors causing the bright band signature for targets in the Rayleigh scattering regime at a precipitation rate of 1 mm/h. The combined effects of the change of the dielectric constant (far left), the change in hydrometeor diameter (middle left), and the change in hydrometeor concentration associated with the change in average hydrometeor fall speeds (middle right) give rise to the bright band signature (far right). Two cases are presented, one assuming a normal snowflake density (solid line), and one assuming a snowflake density five times greater than normal (dotted line) as might occur if snowflakes were to primarily grow by accretion of cloud droplets instead of by diffusion and aggregation.

4.5 Frozen precipitation

Precipitation does not always fall in liquid form. Sometimes it falls as ice, sometimes as melting ice (a type of precipitation that the World Meteorological Organization does not officially recognize but that is very real!). Much variety in form and shape can be observed. For this discussion, we will limit ourselves to the three main types of frozen precipitating hydrometeors: snow, ice pellets (sleet), and hail. They have very different formation mechanisms and hence very different appearance on radar, especially in terms of the vertical structure of echoes.

Since snow is formed by the same phenomena as cold rain, the general structure of its echoes on horizontal sections is similar to that for stratiform cold rain (Fig. 4.9a), except that it generally has less small-scale structure and often looks "fuzzier" than rain on radar displays. Occasionally, behind cold fronts or when cold air is overrunning a warmer surface like an unfrozen water body, it may be convective and assume a more cellular structure (Fig. 4.9b). If the convection is strong enough, snow grains may be observed instead of snowflakes. One of the biggest radar-based forecast challenges associated with snow is to determine when it will start: because of its slow fall speed, snow can both drift over large distances and sublimate for a long time before it reaches the ground (Fig. 4.10). Since we

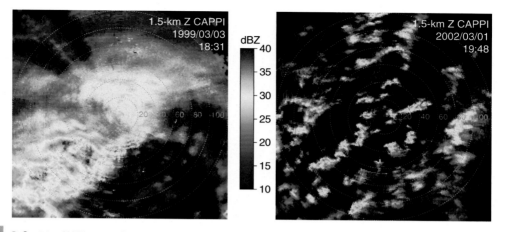

Figure 4.9. Reflectivity CAPPIs at an altitude of 1.5 km in widespread snow ahead of an approaching low-pressure system (left) and in showery snow behind a cold front (right). Range rings are 20 km apart.

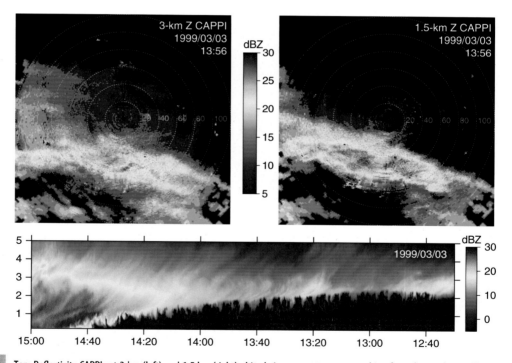

Figure 4.10. Top: Reflectivity CAPPIs at 3 km (left) and 1.5 km (right) altitude in a snowstorm approaching from the southwest. Note how weak echoes at 3 km have progressed further northeast than at 1.5 km. Bottom: Height–time section of reflectivity collected at the scanning radar site. The time axis is reversed so as to have echoes ahead of the system toward the northeast appear on the right side of the image. Several hours can pass before snow aloft reaches the ground. At the base of the echo, the sublimation of snow cools the air more where the snow is more abundant; this produces small local downdrafts that entrain snow down, forming "stalactite" echoes (e.g., Atlas 1955).

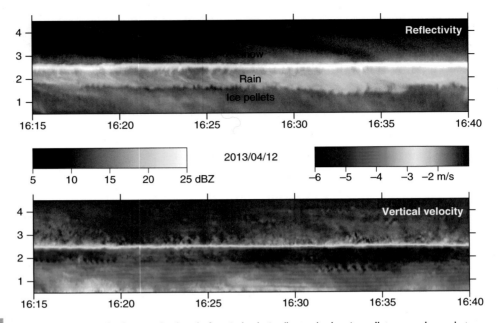

Figure 4.11. Time–height section of reflectivity (top) and of vertical velocity (bottom) when ice pellets were observed at the surface. The melting of snow to rain and its refreezing into ice pellets can be observed.

generally observe echoes at a certain altitude instead of at the surface, we can easily get a wrong impression as to where it is actually snowing on the ground.

Ice pellets form when raindrops, or partially melted snowflakes, freeze as they encounter a deep layer of air at subzero temperatures. Often, the transformation from liquid to solid occurs very close to the ground and remains unobserved by scanning radar. The echoes hence look like (freezing) rain formed by the cold rain process, with a 5-dB loss in reflected power near the ground as raindrops ($|K|^2 \approx 0.93$) change to ice pellets ($|K|^2 \approx 0.18$), also associated with a small change in fall velocity. A rare example where that transformation has occurred high enough to be well observed by a vertically pointing radar is shown in Fig. 4.11. On that image, a small but sharp change in equivalent reflectivity factor can be observed above 1.5 km and is caused by the refreezing of raindrops. Other than for these exceptional cases, the refreezing of raindrops remains difficult to detect by radar in general and scanning radar in particular.

Another common form of frozen precipitation is hail. While ice pellet formation is clearly a stratiform precipitation phenomenon, hail is the result of very deep and severe convection. Hail forms in intense thunderstorms where hail embryos such as graupel are lofted by strong updrafts and grow by capturing supercooled cloud droplets that then freeze on the growing hailstone. Hail echoes will hence typically be in the middle of severe thunderstorm cells, and they used to be recognized only by their very high reflectivity exceeding 55 dBZ. In recent years, hail has become easier to identify using information from radars with dual-polarization capability (Chapter 6).

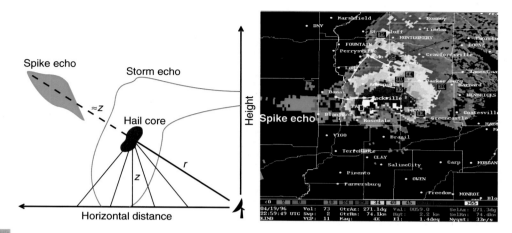

Figure 4.12. Left: Conceptual diagram illustrating how the hail spike echo forms as a result of the reflection of strong hail returns first directed toward the surface, then back to the hail core, and then to the radar. Right: Example from the Indianapolis WSR-88D radar, the radar being located on the right of the image.

An echo "spike" behind the hailstorm (at further range), also referred to as a double-body scattering signature, may occasionally be observed when hail is particularly severe. Figure 4.12 illustrates the phenomenon: when targets are particularly reflective, such as strong hail, multiple reflections may occur before the return echo makes its way back to the radar. These multiple reflections within the hail core result in a delayed return that appears to originate from further ranges than in reality. But most of these remain within the storm, and hence are invisible. Another possible pathway is for reflections to travel to the surface, reflect back to the hail core, and then return to the radar. The resulting echo could then be seen outside of the storm at further ranges and appear as a faint spike echo if a very intense hail core is at a sufficient height above the ground. It is hence a signature associated with severe hail.

4.6 Nonprecipitation weather targets

Not all weather targets are precipitation. Clouds and refractive index gradients can generate detectable returns. Cloud hydrometeors are generally better detected by radars using shorter wavelengths while refractive index gradients become more visible on wind profilers using longer wavelengths.

4.6.1 Clouds

Clouds generate radar returns both because of the presence of hydrometeors and because of sharp refractive index gradients near the cloud top. Clouds can be made of liquid droplets, ice crystals, or a mixture of both.

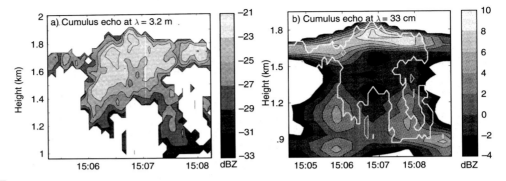

Time–height sections of reflectivity of a cumulus cloud observed by radars operating at 94 GHz (3.2 mm wavelength, left) and at 915 MHz (32.8 cm wavelength, right). To ease comparison, the outline of the echo observed at 94 GHz has been added on the 915 MHz image. Note the change in the reflectivity scale between the two images. While the short-wavelength system is sensitive to cloud droplets, the longer wavelength radar detects clear air echoes caused by the sharp moisture discontinuities at the edges and especially at the top of the boundary layer cumulus cloud. Republished with permission of the AMS from Kollias *et al.* (2001); permission conveyed through Copyright Clearance Center, Inc.

Liquid clouds are made of droplets that measure less than 25 μm. Cloud droplets generally remain small unless a few start to grow via collisions to become drizzle or rain drops. Even though cloud droplets are numerous, their small size makes liquid clouds very difficult to detect by radar; reflectivities are generally below −20 dBZ. For a very long time, it was said that radar was seeing all of the rain but none of the cloud. This is still true for most systems. But with the development of highly sensitive radars working at millimetric wavelengths, it is now possible to observe echoes from cloud droplets (Fig. 4.13).

Ice clouds are made of crystals, generally measured in hundreds of microns. As a result, all ice clouds, even thin cirrus, are much easier to detect by radar than liquid clouds. Most radars can be used to observe ice clouds. Ice crystals generally fall gradually through the cloud at speeds approaching 1 m/s and sublimate at the bottom. Because the growth process is continuous from ice crystals in clouds to precipitating hydrometeors, the distinction between clouds and precipitation is more difficult to make than for liquid precipitation, using both radars and other tools. For example, on the time–height display of Fig. 4.10, one would be challenged to say at what exact point in time or space the echo stops being an ice cloud and becomes instead an echo from solid precipitation.

4.6.2 Refractive index gradients in air

Partial reflection of electromagnetic waves occurs when these waves encounter changes in refractive index. As mentioned in Chapter 2, the refractive index of air in the troposphere varies with pressure, temperature, and humidity. Such gradients can be found everywhere but tend to be stronger at the top of the convective boundary layer (the lower

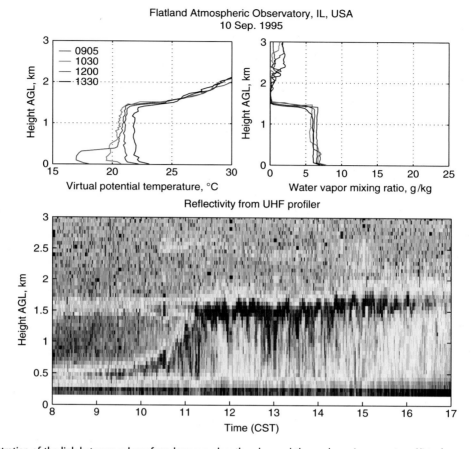

Figure 4.14. Illustration of the link between echoes from long-wavelength radars and thermodynamic parameters. Virtual potential temperature and water vapor mixing ratio from four soundings at the times shown are plotted in the upper panels. The lower panel shows a time–height section of reflectivity from a UHF profiler (arbitrary intensity scale with warmer colors corresponding to stronger echoes). Layers of echoes correspond to levels where the potential temperature and the absolute humidity change rapidly. Republished with permission of the AMS from Angevine *et al.* (1998); permission conveyed through Copyright Clearance Center, Inc.

tropospheric layer in contact with the ground heated by the sun, Fig. 4.14), on frontal surfaces, and on cloud edges (Fig. 4.13). When expressed in reflectivity factor units, returns from refractive index gradients increase as $\lambda^{11/3}$. Clear air echoes are hence primarily observed on radars that are either extremely sensitive or that operate at long wavelengths. They enable wind profilers to work in all weather conditions. While the presence of these echoes is used to detect temperature or humidity boundaries, the intensity information is not used quantitatively at this time to the same extent as the intensity of echoes from hydrometeors.

4.7 Nonweather targets

In addition to echoes of meteorological origin, radars can detect any other target within range. While many of them are considered as clutter, some are meteorologically useful.

In summer, weak echoes from small insects are a common occurrence. They are found throughout the boundary layer, and can hence be used to track winds with reasonable accuracy. Interestingly, they can also be used to determine where low-level convergence and its resulting upward motion occur. The mechanism works as follows: many insects rely on updrafts to be lofted upward, and take advantage of the situation not to have to fly actively, somewhat like a bird gliding on an updraft. Insects within an updraft therefore stay in it for an unusually long time, while more insects are being brought in by low-level convergence. Echo strength from insects is therefore greater in the updraft regions within the atmospheric boundary layer. Identifying areas with stronger insect echoes is extremely useful, as updrafts occur in regions of low-level convergence where new storms may develop. Fronts, dry lines, boundary layer circulations, and many other kinds of low-level convergence regions may hence be revealed thanks to insect echoes (Fig. 4.15).

Birds and large insects like butterflies can also be observed by radar. However, because they move with a significant air velocity, they cannot be used as passive tracers like many insects. Bird echoes come in three main patterns. The first is the point target from an individual bird or a daytime flock; those are generally difficult to spot. A second pattern is

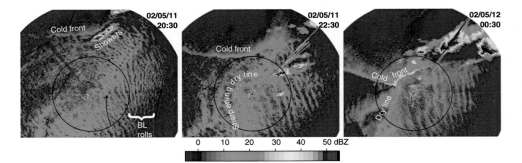

Figure 4.15. Time sequence of low-level PPIs showing echoes from insects (most echoes weaker than 20 dBZ, corresponding to over 95% of the echoes on these images), convective rain (most echoes stronger than 20 dBZ), and a few weak ground targets (especially within 20-km range) collected in western Oklahoma, USA. The circle on the images represents the 60-km range ring. Insect echoes are organized in linear patterns on the upwardly directed branches of boundary layer (BL) circulations, leading to striations in the reflectivity field, and on updrafts forced by larger scale convergence, leading to more intense lines such as the cold front and the dry line. Arrows on the 20:30 UTC image indicate the direction of boundary-layer winds, and major convergence lines revealed by enhanced insect echoes are annotated.

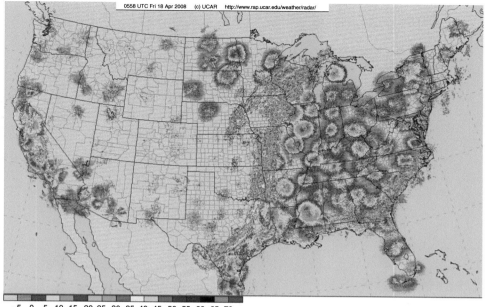

0558 UTC Fri 18 Apr 2008   (c) UCAR   http://www.rap.ucar.edu/weather/radar/

−5 0 5 10 15 20 25 30 35 40 45 50 55 60 65 70
Reflectivity (dBZ)

Figure 4.16. Composite of reflectivity from all conterminous US radars on the night of April 17–18, 2008 (2:00 EDT in the east, 23:00 PDT in the west). One can observe near the center of the continent a relatively strong band of echoes from precipitation associated with a low-pressure system. East of it, disks of weaker echoes caused by bird migration can be seen around almost all radars. In spring, birds migrate primarily ahead of approaching storms where they can benefit from southerly tail winds but not when winds blow in the opposite direction as is the case west of the storm. The reverse is true in fall. Birds may be bird-brained, but they are not stupid! © 2008 UCAR, used with permission.

an expanding disk and ring just after sunrise or sunset as birds get out of their nesting area to start their active day. The most spectacular pattern is the third: bird echoes reaching 20 dBZ covering most of the radar display below 2 km as they migrate at night (Fig. 4.16, supplement e04.1).

Ash and other small debris can be ejected in the atmosphere by fires and volcanic eruptions. Because ash is made of relatively large particles, it can yield fairly strong echoes: the reflectivity of ash from forest fires often reaches 40 dBZ, while that of a volcanic eruption can be much stronger. As seen in Fig. 4.17, ash echoes can be indistinguishable from precipitation on individual images. On vertical cross sections, ash appears to shoot up above the point of origin and slowly drift and fall downwind. That is why, in time animations, the point of origin of ash seems anchored in one place. This is one of the key ways to recognize ash in the absence of other clues such as satellite imagery or information about the presence of a fire. Note however that, on occasion, weather echoes can also appear to originate from only one place, so this clue is not foolproof.

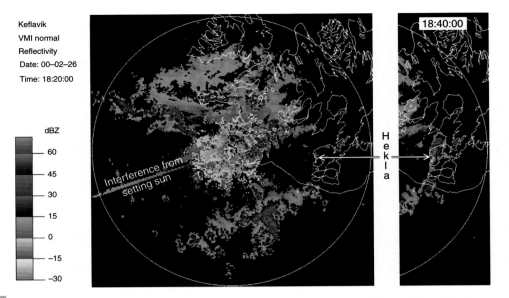

Figure 4.17. Map of maximum intensity in the vertical from the Keflavik radar in Iceland 2 min (left) and 22 min (right) after the beginning of the eruption of the Hekla volcano on February 26, 2000. The location of Hekla is indicated; note that the image on the right only corresponds to the eastern part of the left image. Within 20 min, echoes exceeding 60 dBZ are detected associated with ash from the volcano. Other echoes visible in the two maps include ground and sea clutter at close range, two broad regions of precipitation to the north and south, and interference from the setting sun (along with visible light, the sun emits weakly in the microwaves; when the radar points at the sun, these emissions are detected by the radar and can be wrongly interpreted as "echoes"; other sources such as local area networks (LAN) may give similar signatures). Images courtesy of Sigrún Karlsdóttir using data from the Icelandic Meteorological Office.

Ash is not the only "dry precipitation" whose echoes can be observed by radar. Chaff is another. Chaff fibers are artificial radar targets released in the atmosphere. They are used by the military to create false echoes to fool radar-guided weapons and also for atmospheric dispersion studies. Released in a small volume, they fall slowly drifting with the wind. At first they may look like a suddenly burgeoning storm echo at high levels; but as time progresses, they generally stretch into narrow trails on both the vertical and the horizontal (Fig. 4.18). Depending on the amount of chaff released, their echo can cover a sizable horizontal area (supplement e04.2).

Last but not least, surface targets are by far the most visible nonmeteorological echo. While echoes from water surfaces are generally weak, ground echoes can be extremely strong and are generally very textured (Fig. 4.19). Both types of echoes are very shallow in the sense that their strength decreases much more rapidly with height than is the case with precipitation. Ground echoes can furthermore be recognized by their immobility: after a few days, one learns to identify and mentally filter them. They also have a zero Doppler velocity. On many radars though, ground targets are being removed by the radar

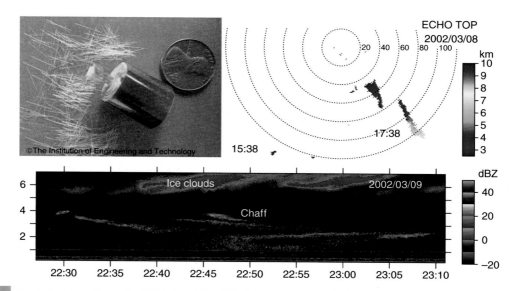

Figure 4.18. Top left: Picture of a small cylindrical container full of chaff fibers, some of which have been pulled out (Stimson 1998, © 1998 IET, used with permission). Chaff is often made of thin glass fibers coated by a reflective metal. Top right: Echo-top map showing chaff soon after its release when it still forms a strong point-like target (15:38), and 2 h later (17:38) when it has taken the shape of a descending trail because of the different fall speeds of each fiber. Bottom: Height–time section of reflectivity of chaff echoes from another release.

signal and data processing systems that use their zero velocity to recognize them and edit them out. That processing is not infallible, and moving ground targets such as wind mills and vehicular traffic can remain, while some precipitation echoes that appear stationary to the radar processing system because they locally have zero Doppler velocity may be removed.

4.8 An echo identification approach

Given the long list of observable targets above, their proper identification may appear to be a daunting task, especially since there can be considerable overlap between the characteristics of different types of echoes. Therefore, there is no easy step-by-step method to determine target type. A good approach is to look at the structure and behavior of the echo and determine whether these fit what would be expected for different types of targets. At the end, most types of targets will be eliminated, and hopefully one type of echo will be found to be the most probable.

Characteristics that should be considered include the following:

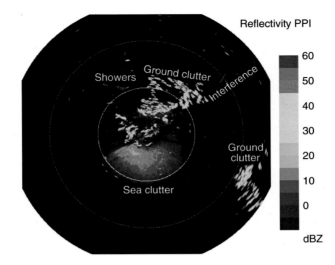

Figure 4.19. PPI of reflectivity at 0.8°PPI showing a variety of unwanted clutter echoes observed by the French Collobrières radar near the Mediterranean Sea. Among a few small shower echoes to the northwest, one can observe echoes from ground targets (strong textured echoes or small arcs along a fixed range), from the sea surface (the lighter more uniform area to the south), and interference from other radar transmitters (short segments along azimuths as well as the southwest–northeast-oriented arc over the "terfe" of "interference" on the image). Many point targets caused by boats and airplanes can also be identified. Range rings are every 100 km. Original image courtesy of Pierre Tabary.

a. Echo strength: What is the average and peak strength of the echo? Near the surface, within the first 2 km, most targets are possible but many have typical echo strength ranges (see also Table 3.2). At usual weather radar wavelengths, strong echoes (>40 dBZ) are from convective precipitation, the strongest (>60 dBZ) being generally associated with hail. Moderate echoes (25–40 dBZ) are generally associated with strong widespread precipitation, but widespread precipitation can have intensities down to below 0 dBZ if it comes from drizzle. Echoes from ash and chaff also have similar intensities. Echoes weaker than 25 dBZ can come from many types of targets, though those from liquid clouds and refractive index gradients are improbable above −10 dBZ. Note that ground targets can have any intensity and therefore fit in all those categories.

b. Size and shape: Targets have a variety of shapes, some of which lead to characteristic echo patterns. Globular echo patterns (Figs. 4.1 right, 4.5 top, and 4.9 right), arranged in clusters or lines of tens to a few hundreds of kilometers, especially if they move, are from precipitation, generally convective. Echoes that are more extended and with less structure (Figs. 4.1 left and 4.9 left), especially if they extend above 3 km, are also from precipitation, generally stratiform. When they are elongated along an axis that is not on a radar radial, they can be from chaff (Fig. 4.18) or from interference if they form an extremely thin line (Fig. 4.19). When elongated along a radial, echoes could also be from a hail spike (only if the echo pattern is short and located immediately behind a very strong target, Fig. 4.12), interference from another source (if it is very thin, extends at all ranges, and remains on the same azimuth with time, Fig. 4.17), or weather echoes that are beyond the maximum unambiguous range of the radar (especially if the pattern moves slowly in the azimuthal direction; see also supplement e05.4). Point targets could be from unedited noise if the reflectivity is weak or from airplanes or isolated birds if the echo is stronger. Note that the perceived size and shape is also affected by the geometry of radar measurements because the altitude of observation and radar sensitivity change with range. As a result, weak or low-level targets whose concentration near the surface is nearly uniform will often appear as disks centered on the radar (Fig. 4.16); this will be the case for birds, insects, sea clutter, and weak stratiform precipitation. Another example is the bright band that will often be detected as a sometimes incomplete ring centered on the radar on PPIs at sufficiently high elevation (Fig. 6.8a).

c. Vertical structure: Echoes extending from the surface to above 3 km are generally from precipitation, ash and chaff remaining a possibility. Within these, the presence of a bright band at an altitude where the 0°C would be expected is a clear signature of precipitation that melted from snow to rain. If the bright band is absent, snow or warm rain is being observed, unless the echo is very strong in which case convective storms are present. Echoes whose intensity decreases rapidly with height from the surface could still have many origins. Strong echoes decreasing very rapidly with height are generally from clutter. Weaker echoes extending to a few kilometers can still be drizzle, snow, insects, or birds. Unusual patterns such as echoes limited to 1 PPI are unlikely to be of meteorological origin and are often caused by interference or the sun.

d. Movement: Movement is a useful clue especially for targets that are not horizontally uniform. This is because weak echoes that have little horizontal structure will appear to be stationary even if they move: the combination of the lack of patterns to track and the inability of radar to detect weak targets beyond a certain range leads to an echo with a disk shape that appears to stay fixed around the radar. An echo pattern that stays absolutely stationary for an extended period (more than 2 h), especially if it has a lot of small-scale structure, is probably from clutter. Echoes that evolve but seem locked in position could be from precipitation that keeps reforming on topography, but are often associated with ash released by a fire or an eruption.

e. External clues: Other data sources can occasionally provide key missing information. In particular, the absence of clouds on satellite imagery rules out all weather targets except for very weak clear air echoes. Temperature is also important to consider: snow is unlikely above 0°C, while birds and insects are unlikely below 0°C; drizzle and rain are also rare below −10°C.

f. Likelihood: When echo characteristics are insufficient to determine the nature of the target, and external clues are ambiguous, using the likelihood of occurrence remains the best approach. Echoes from precipitation are common, and so are ground targets, especially in the absence of clutter filtering. Over land, insect echoes are common when temperatures are warm enough while over oceans, sea clutter occurs regularly at close range if the winds are sufficiently strong. Migrating birds are regularly observed on spring and fall nights, especially if winds are favorable, but not at other times. Interference echoes may or may not be common, depending on the area and the radar frequency. Ash and chaff are generally rare, the occurrence of the latter being a function of how close the radar is located to an air force base.

Even with these tests, there may remain ambiguities, particularly for weaker echoes at low altitudes, and more information is needed. This is where other radar measurements such as Doppler velocity and those derived from multiple polarizations will help.

Doppler velocity information

5.1 Doppler measurements

Doppler radars have been deployed operationally since the 1990s in the United States and are now the norm. In addition to reflectivity, these radars measure the radial velocity of targets, which corresponds to the component of their 3-D velocity moving toward or away from the radar. This information has proven to be critical for helping the detection of severe weather conditions. But even though television meteorologists often insist that the radar imagery being shown is derived from one or several *Doppler* radars, they almost never show velocity imagery. This is because, as we will soon see, Doppler data are more difficult to interpret and at times can be more ambiguous than reflectivity data. In this chapter, the different Doppler quantities that can be measured are introduced, together with the type of information that can be obtained from them.

5.1.1 Doppler moments

Let us consider a point target at range r. In Chapter 2, we saw that the phase φ of such a target with respect to the transmit pulse is

$$\varphi = -2\pi f t_{\text{travel}} = -\frac{4\pi f n}{c} r, \tag{5.1}$$

where f is the radar transmit frequency, t_{travel} is the time required for the radar pulse to reach the target and come back, n is the average refractive index of air along the path, and c is the speed of light in vacuum. If the target range r changes, so does t_{travel} and φ. This change can only occur if the target moves away or toward the radar. The rate of change of φ thus becomes

$$\frac{d\varphi}{dt} = -\frac{4\pi f n}{c} v_{\text{DOP}}, \tag{5.2}$$

where v_{DOP} is the Doppler velocity of the target. If we can measure the rate of change of target phase, we can estimate the radial or Doppler velocity of targets. The mean Doppler velocity is sometimes referred to as the first Doppler moment.

In general, multiple targets are present within the volume sampled by the radar, and they do not all have the same velocity. As a result, the power and Doppler velocity measured from the sampling volume fluctuate because of the changing interference between the returns coming from each target. The rate at which these fluctuations occur is therefore a measure of the spread in target Doppler velocities within the sampling volume. Hence, in addition to a mean Doppler velocity v_{DOP}, the spectrum width σ_v of target velocities can also be measured. Spectrum width is also known as the second Doppler moment. By extension, reflectivity is sometimes referred to as the zeroth Doppler moment. An example of data from all three Doppler moments is shown in Fig. 5.1. Unless the beam width of the radar is extremely narrow, the quantitative interpretation of spectrum width is often difficult because of the many processes that can affect it (Fig. 5.2). As a result, it has not been widely used compared with the mean Doppler velocity, though it is still useful qualitatively to identify regions with high wind variability.

The sign convention for Doppler velocity is not consistent for all radars. Most scanning radars use the "PANT" convention, for positive away, negative toward. Consequently, when the radar points up, positive numbers correspond to upward velocities. However, many wind profilers use the opposite convention. But there are exceptions in both cases, so beware! What everyone more or less agrees on is to use cold colors such as blues and greens for velocities moving toward the radar, and warm colors such as yellows and reds for velocities moving away from the radar.

5.1.2 Doppler spectra

As seen in Chapter 3, the signal that comes back to the radar is the sum of the returns from all targets. A spectral decomposition of the signal can be made by computing the Fourier transform and then the power spectrum of the raw signal at a given range, thus obtaining the distribution of the total power as a function of Doppler velocity. The resulting function is known as the Doppler spectrum of echoes at a given range. Increasingly, radar signal processors compute Doppler spectra as a first step toward data cleaning such as ground echo suppression. These spectra are not often archived because of the large amount of media space required and of the rare scientific use made of them. The only exception is with vertically pointing radars and wind profilers, when the Doppler spectra can be interpreted as distributions of target fall speeds as a function of height. The information so retrieved becomes considerably richer (Fig. 5.3).

5.2 Information at vertical incidence and profiling

When pointing either up from the ground or down from an airplane or a satellite, Doppler radars measure the vertical velocity of targets. In spite of the limitations of taking measurements in only two dimensions (height z and time t) as opposed to four (x, y, z, t) for scanning radars, the detailed information derived from such data is extremely valuable and not fully exploited at this time.

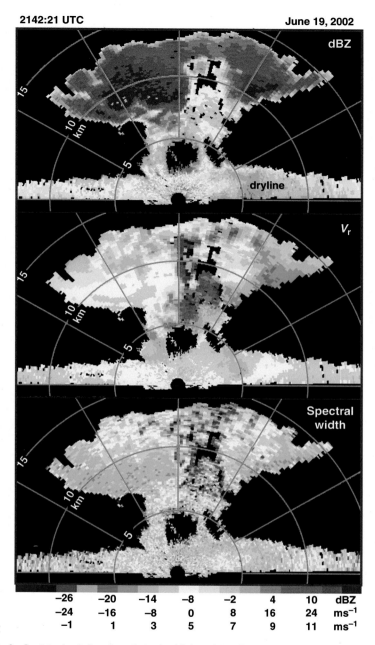

2142:21 UTC   **June 19, 2002**

−26	−20	−14	−8	−2	4	10	dBZ
−24	−16	−8	0	8	16	24	ms⁻¹
−1	1	3	5	7	9	11	ms⁻¹

Figure 5.1. Conical scan of reflectivity (top), Doppler velocity (middle), and Doppler spectrum width (bottom) made by an airborne radar through a growing cumulonimbus cloud in NW Kansas, USA. In these images, the left–right direction is in the horizontal while the up–down direction is close to the vertical. Numbers associated with the color scale below are for reflectivity (first line), Doppler velocity (second line), and spectrum width (third line). Little precipitation has been formed yet given the weak reflectivity values observed. On the Doppler image, positive velocities just to the right of the center of the image indicate the location of the main updraft. The updraft may be more turbulent, as suggested by the higher spectrum widths, while the downdrafts appear to be steadier. Republished with permission of the AMS from Wakimoto *et al.* (2004); permission conveyed through Copyright Clearance Center, Inc.

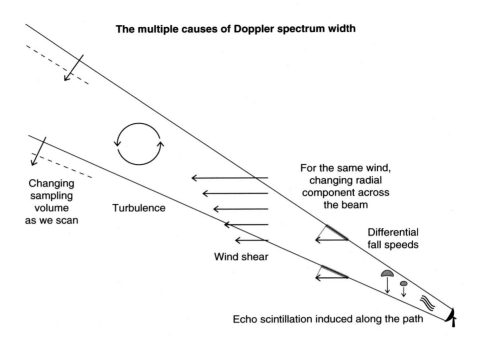

The multiple causes of Doppler spectrum width

Changing sampling volume as we scan

Turbulence

Wind shear

For the same wind, changing radial component across the beam

Differential fall speeds

Echo scintillation induced along the path

Figure 5.2. Illustration of the multiple processes that broaden the width of the Doppler spectrum. These arise from the fact that either between one edge of the beam to another, or between the beginning and the end of the measurement time, the same phenomena will lead to different Doppler velocities.

Velocity measurements at vertical incidence depend both on the type of targets observed and on the dynamics of the atmosphere. In Chapter 4, we discussed how to recognize the many different types of targets that are observed with weather radars. The task of identifying targets becomes much easier when their fall velocity is known. The classification can also be made with more detail (Fig. 5.4). For example, rain with bigger drops can be distinguished from rain with smaller drops, an important consideration for converting reflectivity to rainfall rate. Intense rain is often the result of larger drops that fall faster; hence, rain with high reflectivity should have faster fall velocity than rain with low reflectivity or drizzle. This is generally true, but it is not always the case. In Fig. 5.4, one can see such an example by contrasting the faster falling rain around 1:10 (under the term "Rain") and the slower falling rain at 1:25 that has an identical reflectivity. Because rain drops of a given size fall at a predictable velocity, Doppler spectra measurements can be used to retrieve the distribution of drop sizes, provided that the vertical velocity of air is known or negligible. Moreover, Doppler spectra measurements can be used to verify whether there is one or multiple target types within the same sampling volume (Figs. 4.4 and 5.3).

But the most common use of Doppler information at vertical incidence has been to monitor the dynamics of the atmosphere at the convective and cloud scales. This is beautifully illustrated in Fig. 5.5: in this single 30-min image, one can observe boundary layer motions, the lifting of warm air by a cold outflow and how it is linked with enhanced insect echoes, cumulus clouds at various stages of evolution, and the complex interactions

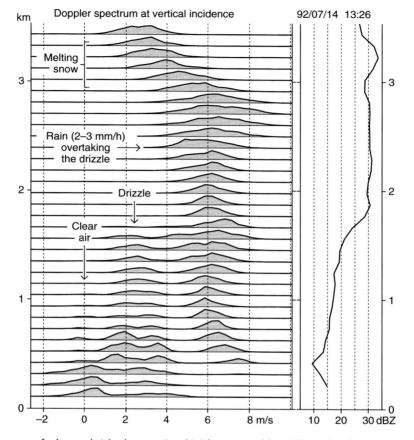

Doppler spectra of echoes at heights between 0 and 3.5 km measured by a UHF wind profiler at 13:26 EDT on July 14, 1992, in Montreal, Canada. In the main window (left), each of the 33 curves shows the relative contribution to the reflectivity of targets moving at a given vertical velocity (downward when positive) as a function of height. The vertical profile of reflectivity is plotted on the right. Three types of echoes can be identified: clear air echoes (0–1 km height) with velocities near 0; drizzle (0–1.6 km height) with fall velocities centered around 2.5 m/s; stratiform precipitation (0.5–3.5 km) in the form of snow at 3.5 km and rain below 3 km, with melting snow in between. At this instant, stratiform rain was just starting and was still mainly above 1.8 km, masking the drizzle echoes. Only a few drops had already reached the lower altitudes, with the largest (fastest, nearly 8 m/s) ones being the closest to the surface. From Fabry *et al.* (1993), © Copyright 1993 AMS.

between all these processes. An entire lecture would be required to properly explain what is being observed during this short time interval!

At vertical incidence, the interpretation of Doppler data for the purpose of obtaining information on air motion is both simpler and more complicated than for scanning radars. On the one hand, the geometry of the measurement is simple: the information obtained is the vertical velocity of the targets. On the other hand, that information is a combination of a nonnegligible fall velocity of targets with respect to still air, and of the vertical velocity of

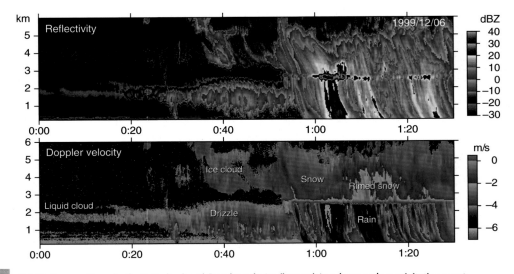

Height–time section of reflectivity (top) and Doppler velocity (bottom) in a large-scale precipitation event. Many types of weather targets can be observed and identified.

the air. For liquid cloud echoes such as (7a)–(7e) in Fig. 5.5, this is not an issue. For all other targets, the vertical velocity of targets with respect to still air must be considered. Insects, for example, both actively move and are carried away by vertical motions. In Fig. 5.6, we can see them do both. During the daytime hours, insects are pushed up and down by air motions much like they seemed to be in Fig. 5.5. But one can clearly observe different types of insects actively taking off just after sunset and just before sunrise; and they have to land at some point. Indeed, Geerts and Miao (2005) showed using aircraft data that most of the time, insect echoes move down with respect to air motion. However, as long as the information is used more qualitatively than quantitatively, the motion of insect echoes can be a powerful revelator of atmospheric motions.

The same is sometimes true of snow echoes, especially ahead of low-pressure systems: if the air motions are sufficiently small in intensity or in scale to have limited effects on the density and resulting fall speed of snowflakes with respect to still air, snow Doppler velocity will mimic air velocity. In these events, one can observe both wave motions arising in convectively stable layers (Fig. 5.7) and small convective cells generated by the horizontally inhomogeneous latent heat transfer that occurs when ice deposits, sublimates, and melts.

Under most other circumstances, target fall velocity and air motion interact: updrafts bring moisture that leads to precipitation growth and a change in fall velocity. For most rain events, vertical velocity is more a function of drop size than air motion. The direct evidence of air motion is rarely seen in such images (e.g., Fig. 5.4), even though we know it has to be present as it shaped precipitation growth rates and processes. Only in severe convection will air motion be strong enough to leave an identifiable mark on Doppler velocity (Figs. 4.3 and 5.1).

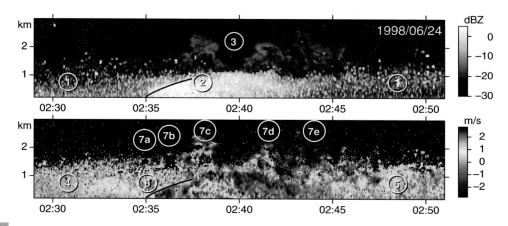

Figure 5.5. Time–height image of reflectivity (top) and Doppler velocity (bottom) from an X-band vertically pointing radar during the passage of an outflow boundary originating from a distant thunderstorm. On the Doppler image, positive velocities and warm colors correspond to upward motion. Using reflectivity, one can observe insect echoes (1), enhanced insect echoes along the boundary (2), and cumulus clouds (3). With velocity information, the picture becomes more detailed. Ahead of the cold outflow (4), the air is calm in the nighttime boundary layer as seen by the uniform velocity field. But behind it (5), it is much more turbulent. Convergence at the boundary lifts air as well as insects (6). This lifting forces air parcels through the capping inversion of the boundary layer, resulting in the formation of cumulus clouds. Five cumulus clouds can be observed at different stages of their evolution, the more mature clouds being the furthest away from the convergence line. The first two cumuli (7a and 7b) are mere hot bubbles that are still within the low-level echoes caused by insects. The next one (7c) is a vigorous cumulus cloud, with an updraft in its core and downdrafts at the periphery. The fourth cloud (7d) is at an early stage of dissipation with a few updrafts and many downdrafts, while the last one (7e) is collapsing rapidly. New storms formed over this outflow boundary 20 min later. Reprinted from Fabry and Zawadzki (2001), © Copyright 2001 with permission from Elsevier.

Last but not least, two important applications of Doppler velocity at near-vertical incidence are wind profiling and radio acoustic sounding. Wind profilers are long-wavelength radars designed specifically to measure the vertical profile of wind above the radar as a function of time by using the echoes from precipitation as well as from clear air. Most wind profilers work as follows (Fig. 5.8): first, the radar points vertically for a period of about 30 s to obtain data on the vertical velocity of targets. Figure 5.3 showed an example of such data. Then the radar points at a high elevation θ such as 75° toward one cardinal direction, say the East, for a similar time period. The Doppler signal received will hence be from a combination of the vertical wind w and the east–west horizontal wind component u, that is, $w \sin(\theta) + u \cos(\theta)$. Since w is known from the measurement at vertical incidence, u can be derived. The profiler then points toward another cardinal direction, say the North, and from that measurement the north–south component of the wind v is retrieved. Often, profilers will also make measurements toward the South and the West. Because the sampling regions where these three or five measurements are made are different, many measurements must be averaged to eliminate the noise in the wind retrieval. Typically,

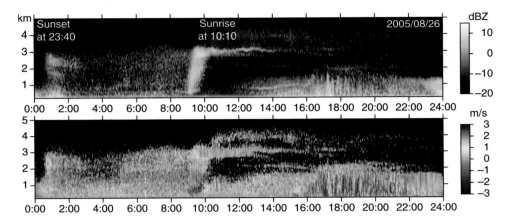

Figure 5.6. Height–time section of reflectivity (top) and Doppler velocity (bottom) during a clear summer day in Montreal, Canada. Echoes in this image are primarily from insects. Using the velocity and reflectivity information, we can clearly observe that many insects take off an hour after sunset and 1 h before sunrise, and gradually come back down later on. This partly explains why the coverage of insect echoes suddenly expands on scanning radar displays at these times. As we move toward midday and the afternoon (16:00–22:00 UTC in Montreal), insects are being increasingly carried vertically by air circulations in the convective boundary layer.

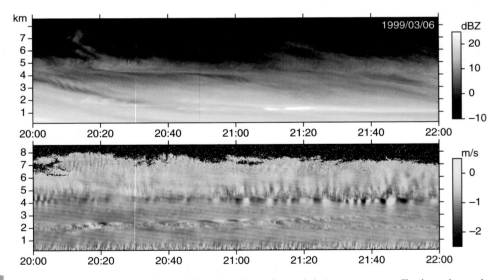

Figure 5.7. Time–height image of reflectivity (top) and Doppler velocity (bottom) during a snowstorm. Thanks to the nearly uniform fall velocity and small inertia of snowflakes, we are able to observe on Doppler velocity the detailed structure of vertical air motions occurring at the time. In this unusually rich example, one can identify mechanical turbulence from surface friction (below 0.5 km), thermally induced circulations in generating cells (e.g., above 7 km at 20:10), and several layers of waves in regions of high wind shear and strong static stability.

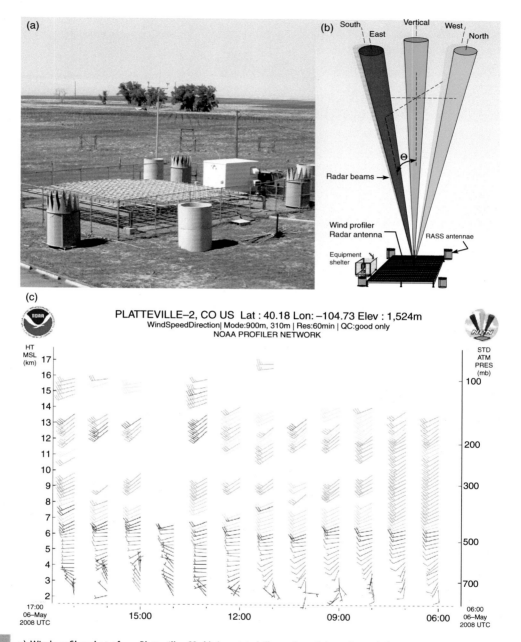

Figure 5.8. a) Wind profiler photo from Platteville, CO. b) Annotated illustration of the radar and the measurement process whereby successive measurements in three or more directions are used to retrieve wind profile information. c) Example of measured vertical wind profiles with height as a function of time. Images and photos are courtesy of NOAA.

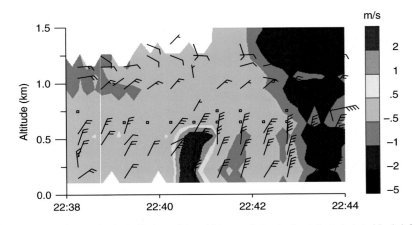

Figure 5.9. Time–height section of vertical velocity (color scale) and 30-s resolution horizontal winds (wind barbs) from an interferometric UHF profiler in Erie, Colorado, USA, in clear air (before 22:43) and snow (from 22:43). The passage of a cold front at 22:41 results in a sudden upward motion and an acceleration of winds from 20 to 30 kt at low levels. Snow with a fall speed of more than 1 m/s follows 2 min later. Image courtesy of William Brown.

profilers can retrieve accurate wind profiles every 30 min. Higher resolution data are occasionally plotted but should not be fully trusted on traditional wind profilers. Since the beginning of this century, profilers using what are known as spaced antenna techniques have been developed so that all components of the wind can be obtained from the same sampling volume. With such profilers, winds at 30-s resolution can now be obtained (Fig. 5.9).

In RASS, a wind profiler is coupled with an acoustic source originating from large speakers (see Fig. 5.8a and Fig. 5.8b). In this unusual setup, the role of the speakers is to generate sound waves that will have half the wavelength of the profiler radio waves. Sound being a pressure wave, and because the refractive index of air is a function of pressure, the acoustic wave sets up a regular refractive index pattern of the proper wavelength to reflect the profiler radio waves. These sound waves hence become an intentional target that travels upward. The main interest of this idea is that, as a result, the speed of sound as a function of height can be measured. The speed of sound c_s is a function of the square root of the virtual temperature T_v of air:

$$c_s = \sqrt{\frac{c_p}{c_v} R' T_v} = \sqrt{\frac{c_p}{c_v} R' T \left(\frac{1 + r_v/\varepsilon}{1 + r_v} \right)}, \tag{5.3}$$

where c_p and c_v are the specific heats of air at constant pressure and constant volume, respectively ($c_p/c_v \approx 1.4$), and R' is the gas constant of air (287 J/(K kg)). Recall that virtual temperature is the temperature at which the observed moist air at temperature T and dry air at virtual temperature T_v have the same density, and is a function of the mixing ratio of water vapor r_v, corrected by ε, the ratio of the gas constants of air and water vapor (≈ 0.622). As a result, from speed of sound measurements, vertical profiles of T_v can be obtained

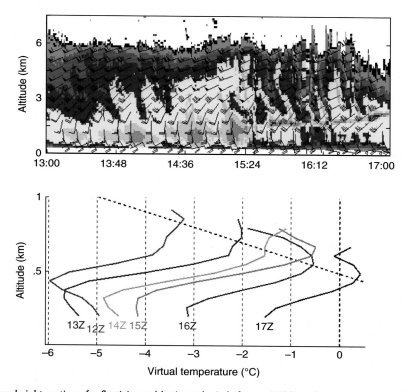

Figure 5.10. Top: Time–height section of reflectivity and horizontal winds from a UHF boundary layer wind profiler in Montreal, Canada, during an approaching warm front in winter. Warmer colors correspond to stronger reflectivity. Snow prior to 15:30 changes to rain afterward, and a bright band appears at around 1.5 km. Bottom: Associated virtual temperature profiles measured hourly by RASS from 12:00 to 17:00. The slanted dashed line corresponds to the dry adiabatic lapse rate. One can observe both warming surface temperatures and the descent of the warm frontal surface. The warmest air around 1.5–2 km is, however, beyond the range of the RASS at UHF frequencies.

(Fig. 5.10). Because both T and T_v are close to each other, it is common to think of RASS profiles as measurements of temperature.

The range up to which virtual temperature can be measured depends on how far the sound travels coherently. As you probably noticed, low-frequency sounds, like the rumble of distant thunder, travel further than higher frequency sounds. Hence, lower frequency profilers, that require lower frequency sounds, will be able to profile virtual temperature higher in the atmosphere: higher frequency boundary layer profilers will be able to measure temperatures up to about 1 km, while lower frequency profilers can make these measurements up to much greater distances provided the neighbors do not complain about the loud, low-frequency sound emitted by the RASS speakers!

5.3 Mean Doppler velocity on PPI displays

Most Doppler radars scan in azimuth. The mean Doppler velocity measured by a scanning radar is hence the radial velocity, the component of the wind moving either toward or away from the radar. Because the geometry of the measurement changes as the radar scans, Doppler patterns are not as straightforward as those at vertical incidence. Assuming a constant wind from the west, strong approaching velocities would be observed when the radar points toward the west, strong receding velocities when the radar points toward the east, and no velocity when the radar looks north or south (Fig. 5.11). Radial velocity images are usually more complicated than in this simple example because

a. the wind is rarely uniform;
b. the area over which wind information can be obtained is limited to regions with targets (like rain, bugs, etc.) because the measurement of the radial velocity is possible only if there is a target reflecting radar waves and whose speed can then be measured;
c. not all targets move with the wind: think of ground and sea clutter, birds, etc.;
d. the height at which the radar observes targets increases with distance.

Interpreting radial velocity patterns under such conditions requires both experience and the use of some simplifying assumptions to grasp the information presented. Experience comes with practice. As for simplifications, several idealized models are used depending on whether the goal is to obtain wind profiles with height, the horizontal wind patterns at the

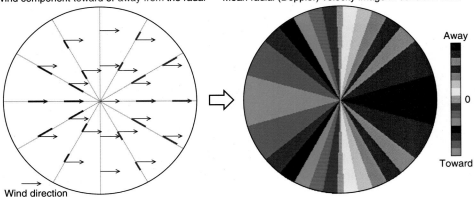

Figure 5.11. Illustration of the Doppler measurements obtained given a spatially uniform wind from the west. The left image shows the wind vectors as arrows as well as the projection of these wind vectors on radials coming from the radar. For each vector, the length of thick lines shows the magnitude of the radial velocity measured, blue being used for radial velocities toward the radar and red for radial velocities away from the radar. If there are echoes everywhere, the resulting mean Doppler velocity measured is shown on the right.

large-scale end of the mesoscale, or convective-scale signatures associated with severe weather.

5.3.1 Two paradigms: measuring on a plane vs. on a cone

The most commonly used Doppler image is from a PPI. As a result, as range increases, so does the height sampled by the radar beam (recall Fig. 2.18). Measurements made at close range sample winds at lower heights, while measurements at far ranges sample winds at higher heights, a simple fact that is essential to keep in mind for a proper interpretation of Doppler imagery.

Depending on the elevation of the PPI scan and on the type of wind patterns expected, two different conceptual approaches can be used to simplify the interpretation of Doppler imagery. One assumption is that the wind $\mathbf{v}(x, y, z)$ is sufficiently constant over the ranges of heights sampled by the PPI that the effect of the changing altitude of the PPI is negligible, that is, $\mathbf{v}(x, y, z) \approx \mathbf{v}(x, y)$. In such a case, the structure of the mean Doppler velocity map is dominated by mesoscale and convective-scale changes in winds such as fronts or storm-induced circulations. This way of thinking is most appropriate for near-horizontal PPIs made in summerlike conditions, particularly when severe weather is expected to change winds considerably over short distances. This remains a difficult interpretation to make, because of the need to retrieve $\mathbf{v}(x, y)$, a two-component wind vector varying in two dimensions, using radial velocity, a one-component quantity. Except for a few characteristic signatures, Doppler maps are extremely hard to interpret in terms of a two-component wind vector.

The second approach is to assume that the wind $\mathbf{v}(x, y, z)$ varies with height much more rapidly than it varies with horizontal space, that is, $\mathbf{v}(x, y, z) \approx \mathbf{v}(z)$. If this assumption is correct, since height is a function of range on PPIs, wind speed and direction should hence be only a function of range on PPIs. At a given range or height, the radial velocity measurements at all azimuths can be used to obtain $\mathbf{v}(z)$. Although such Doppler images can be very intimidating, with images like the yin–yang pattern of the Korean flag, they are actually much easier to interpret than when winds vary considerably in the horizontal.

Which approach should be chosen in any particular circumstance? If wind direction and velocity do not change significantly over the area of radar coverage, the Doppler pattern should show perfect central symmetry if one neglects the contribution to the Doppler signal of hydrometeor fall speeds, with

$$v_{\mathrm{DOP}}(r, \phi) = -v_{\mathrm{DOP}}(r, \phi + 180°). \tag{5.4}$$

This should be expected in regions of widespread precipitation, and the second approach works best under this scenario. If this symmetry is broken, it implies that winds also vary horizontally, and the first approach may be a better one, provided that the elevation angle is small. Otherwise, a blend of the two is required, and proper interpretation of Doppler patterns can become very difficult.

5.3.2 Unraveling Doppler's twists and turns

Figure 5.12 illustrates a procedure for extracting a wind profile from a Doppler PPI. Given the possible complexity of Doppler patterns, the following systematic but robust approach is suggested:

1. For the altitude of interest, 3.4 km in this example, identify the data that are at that altitude. In a PPI, these data are found within a ring at constant range, in this example at 60 km. It is important to mentally filter out all other data at different altitudes; otherwise, their use will almost always lead to mistakes. Consequently, all data not within the range of interest have been masked on the right side of Fig. 5.12.
2. Locate where the Doppler velocity is zero. With a simple wind pattern and complete data coverage, there should be two such locations opposite to each other with respect to the radar position. This is not necessarily the case with a more complex spatial wind structure or if echo coverage is incomplete. The axis linking the radar to these zero-velocity regions is perpendicular to the wind at the location where zero velocity is observed. The axis between both zero-velocity regions should be roughly perpendicular to the wind over the radar site at that altitude.
3. Follow the ring until you find the peak negative Doppler velocity (as in Fig. 5.12) and/or the peak positive Doppler velocities.
4. Peak velocities should be observed roughly 90° away from the zero-velocity regions. The wind then blows directly toward the radar from the direction of the peak negative velocity (label "4") or away from the radar to the direction of the peak positive Doppler velocity (180° away from "4," in this case to the northwest). Wind speed is hence the

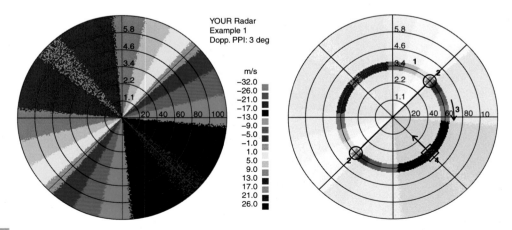

Figure 5.12. Left: Hypothetical Doppler pattern on a 3° PPI associated with an unknown wind profile. Range rings are labeled with their distance from the radar (along the horizontal radial) and their height with respect to the radar (along the vertical radial). Right: The order of the steps needed to determine the wind at a particular altitude. Details are in the text.

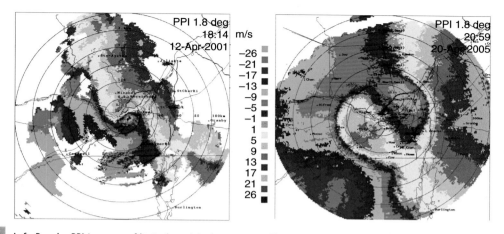

Figure 5.13. Left: Doppler PPI in a case of limited precipitation coverage. The zero-velocity region 60 km south of the radar is due to ground targets. Right: Doppler PPI in a situation where winds at 3.5 km (around 90-km range) are changing direction over the radar coverage, as illustrated by the fact that peak "toward" velocities are not observed 180° away from the peak "away" velocities.

magnitude of that peak velocity. In the example in Fig. 5.12, the wind at 3.4 km is 17 m/s from the southeast. In this specific example, a similar result will be found if the process is repeated at other altitudes.

At first glance, there seems to be a certain amount of redundancy in the method proposed. But it will prove to be helpful in situations with missing or somewhat contradictory information such as when the wind is not uniform in space at the height of interest (Fig. 5.13): flow can be deformed by dynamics or changing topography, leading to curvature, confluence, sloping jets, and a variety of mesoscale and convective-scale circulations, to which one must add contamination by ground echoes. The redundancy of the suggested approach is more likely to help resolve the inconsistencies, thus enabling the retrieval of the average wind profile.

There is only one way to master the interpretation of Doppler velocity patterns: practice. I strongly recommend you now use the procedure described above in the tutorial provided on the electronic material (supplement e05.1) before proceeding further. And yes, this recommendation applies to you too! See you in an hour . . .

Real winds do change with space and in time, even in large-scale systems. Often, the change is gradual (supplement e05.2), but it can be quite sudden in the case of a front. In the example shown in Fig. 5.14, one can see three regimes: a pre-frontal area to the east, followed by a transition zone, which is itself followed by the post-frontal area to the west. Both pre-frontal and post-frontal regions have different but clear wind flows; in order to determine these two wind flows, all other data including that of the transition zone must be masked while imagining what the Doppler field would have looked like had the front not been there. Only then can the winds on both sides of the front be

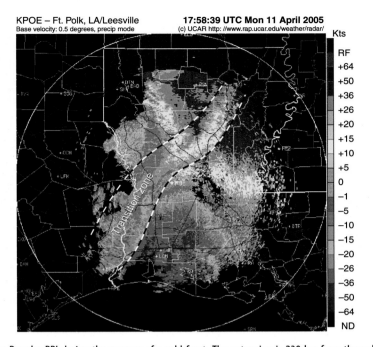

KPOE – Ft. Polk, LA/Leesville   17:58:39 UTC Mon 11 April 2005
Base velocity: 0.5 degrees, precip mode   (c) UCAR http://www.rap.ucar.edu/weather/radar/

Figure 5.14. Low-elevation Doppler PPI during the passage of a cold front. The outer ring is 230 km from the radar. Note the change in wind direction to the east and the west of the transition zone. Ignore the area in purple that denotes echoes for which a Doppler velocity measurement cannot be obtained. © 2005 UCAR, used with permission.

properly determined. The transition zone is more complex; this is partly due to convective-scale circulations occurring within it. We will come back to it later.

5.3.3 Local circulations

By their nature, convective storms have powerful wind circulations that have much smaller scales than midlatitude cyclones. In such cases, one can find patterns in the Doppler imagery associated with these complex circulations. Some of those can be recognized, though not entirely without ambiguity.

To illustrate how these flows would appear on Doppler images, first consider the westerly wind observations from Fig. 5.11. To the mean wind and its associated Doppler pattern, local circulations that are expected in convective systems are added. For this exercise, we suppose that these circulations occur in a small square area to the north of the radar. In this zone, the mean Doppler velocity is as in Fig. 5.15c.

A first possible signature is one of rotation around a point. If there is a rotating circulation, a couplet of "toward" and "away" velocities will be observed side by side at the same range from the radar (represented in blues and purples, respectively, in Fig. 5.15).

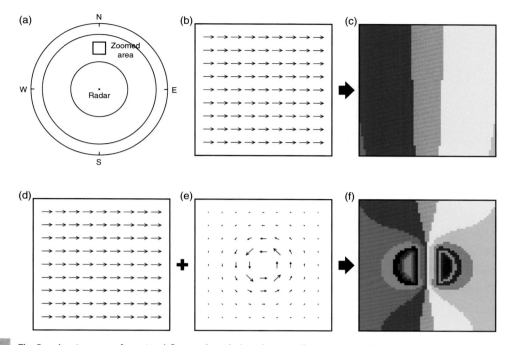

Figure 5.15. The Doppler signature of rotational flow explained. Consider a small region north of the radar (a) where winds are from the west (b) giving the resulting Doppler pattern (c). Add to that wind (d) a rotational flow (e) and the Doppler image will show a couplet of away and toward velocities side by side (f) at the same range. Inspired by Brown and Wood (2006).

This is sometimes referred to as a "kidney signature." The tangential part of the circulation remains unobserved as usual.

There are three main phenomena that can cause rotating wind circulation signatures on Doppler radars. First, the center of a tropical cyclone has a rotating circulation that is small enough to fit within the coverage of a single radar, giving rise to the appearance of large couplets several tens of kilometers wide (Fig. 8.12). Second, at smaller scales, mesocyclones are the most commonly observed rotating wind circulations observed by radar. Measuring a few kilometers in diameter, they are found in powerful convective storms, generally caused by rotating updrafts or downdrafts. They are often the birthplace of tornadoes, hence the importance of recognizing them. For example, three mesocyclones can be seen at the top of Fig. 5.16 and in supplement e05.3. Finally, at an even smaller scale of the order of a few hundred meters, tornadoes will also show a similar signature but only if they are extremely close to the radar (bottom of Fig. 5.16); usually, these tornado vortex signatures (TVSs) are difficult to observe because of their small scale.

Another local Doppler signature observable by Doppler radars and associated with convection is the divergence signature. A divergent circulation is recognized by a couplet of

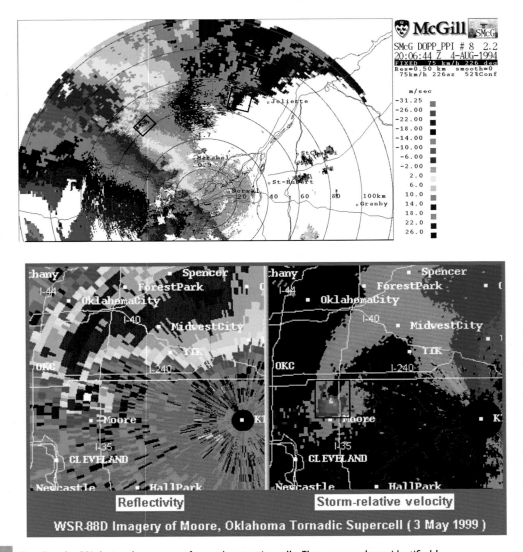

Figure 5.16. Top: Doppler PPI during the passage of several convective cells. Three mesocyclones identified by rectangles can be clearly observed, one 70 km to the northwest and two 70 and 90 km to the north-northeast of the radar. Bottom: Reflectivity and Doppler PPIs of a severe storm that resulted in the 1999 Moore tornado observed here 20 km west of the radar. On the reflectivity image, warmer colors (yellows, reds, and white) correspond to stronger reflectivity; on the Doppler images, greens correspond to velocities toward the radar and reds away from the radar (bottom image courtesy of NOAA/SPC).

"toward" and "away" velocities aligned along the same azimuth, the "toward" velocities being closer in range (Fig. 5.17). The opposite arrangement is observed in a convergent flow.

Keep in mind that these interpretations are only valid for relatively small-scale circulations where the signatures are not centered at the radar location. Strong

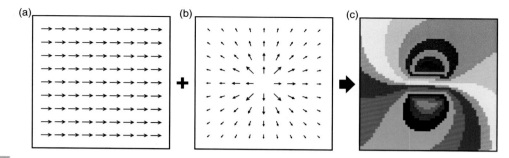

Figure 5.17. The Doppler signature of divergent flow explained. Consider a small region north of the radar where winds are from the west (a). Add to that wind a divergent flow (b) and the Doppler image will show a couplet of away and toward velocities one behind the other along the same azimuth (c). The couplet will be reversed in the case of convergent flow. Inspired by Brown and Wood (2006).

couplets of toward–away velocities can also be observed centered on the radar location, but these are associated with wind maxima at a particular altitude, such as low-level jets.

Convergent and divergent flows occur in a variety of situations in severe convective storms. Of particular importance is the downburst, a strong convectively generated downdraft that expands radially outward as it reaches the ground. Downbursts have been responsible for many airplane accidents on landing, and much effort has been made to ensure their detection as early as possible. On Doppler radar, downbursts can be detected when they hit the ground because of the flow divergence near the surface (Fig. 5.18). But such signatures are very shallow (Fig. 5.19), and can only be detected with low-level scans and at short-to-medium ranges.

Divergence signatures can also extend over larger areas if they originate from multiple downdrafts, as would be the case with a line of showers or thundershowers. This is exactly what happened in the transition region of Fig. 5.14, and on another front passage shown in Fig. 5.20: the cold front caused a line of showers and thundershowers, each with an associated downdraft. The flow from each downdraft diverges as it reaches the surface; these flows then merge to create an outflow originating from an axis close to that of the strongest precipitation. Larger scale convergence lines can also be seen associated with other circulations such as sea breezes or merging gust fronts from multiple thunderstorms. In such cases, in the absence of precipitation, reflectivity fine lines (Fig. 4.15) collocated with convergence signatures on Doppler imagery are often seen.

5.4 Velocity-based products

In the same way that we have developed algorithms for extracting information from reflectivity data, we have at our disposal a variety of products based on Doppler

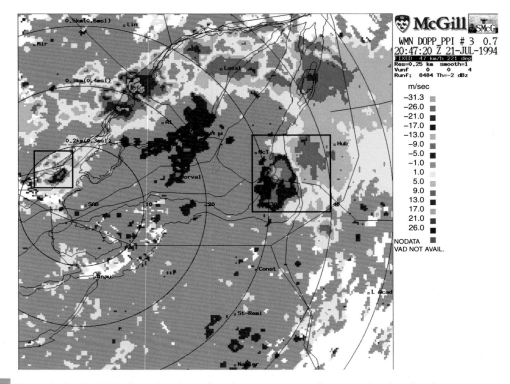

Figure 5.18. Zoom of a low-level PPI of Doppler velocity done during a severe weather event. Two clear downburst signatures identified by squares can be observed.

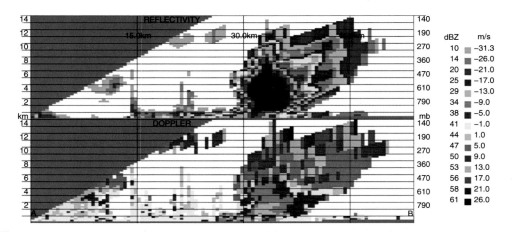

Figure 5.19. Vertical cross section of reflectivity and Doppler velocity passing through a small hailstorm. The radar is to the left, and negative velocities go in that direction. A very shallow divergence signature (green–red couplet on Doppler imagery) can be seen below 1 km between 30 and 34 km in range.

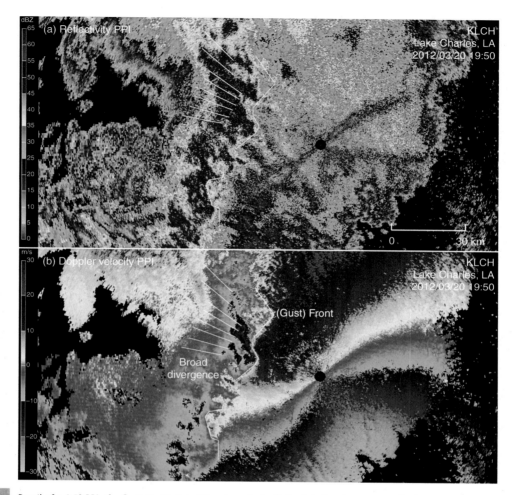

Figure 5.20. Detail of a 0.5° PPI of reflectivity (top) and Doppler velocity (below) during the passage of a front in Louisiana on March 20, 2012. The radar is located on the red lozenge within the small black disk. Annotated on both images are the positions of a band of divergence underneath the precipitation core caused by downdrafts hitting the ground, and of a sharp convergence boundary where the cool downdraft air meets and lifts the pre-frontal air to the east, forming a gust front.

measurements. As with reflectivity-based products, some of the Doppler products are designed to identify severe weather signatures and warn the forecaster accordingly. Two examples are algorithms designed to detect the mesocyclone and near-surface divergence signatures we have just seen. Another very common Doppler product is referred to as the velocity–azimuth display (VAD). The VAD algorithm attempts to retrieve the vertical profile of winds by fitting a function to the Doppler data at each range. Basically, the VAD algorithm performs in a more systematic way the steps we have outlined earlier for retrieving the average wind over the area covered by the radar (e.g., Fig. 5.12). Given a uniform wind profile over the radar coverage, the measured Doppler velocity can be expressed as follows:

$$v_{DOP}[z(r), \phi] = [u(z) \sin \phi + v(z) \cos \phi] \cos \theta' + [w(z) - w_f] \sin \theta' \qquad (5.5)$$

where z is height, ϕ is the azimuth, and θ' is the elevation angle of the radar ray at range r, (u, v, w) are the (east–west, north–south, up–down) components of the wind, and w_f is the fall speed of the targets with respect to air, defined to be positive downward. Equation (5.5) can be rewritten as

$$v_{DOP}[z(r), \phi] = [w(z) - w_f] \sin \theta' + [u(z) \cos \theta'] \sin \phi + [v(z) \cos \theta'] \cos \phi. \qquad (5.6)$$

In (5.6), the first of three terms in brackets on the right-hand side is independent of azimuth, the second is a multiplicative "constant" to the sine of azimuth, and the third is another multiplicative "constant" to the cosine of azimuth. By fitting a function $f(\phi) = a_0 + a_1 \sin \phi + b_1 \cos \phi$ to the Doppler data at a given range, the horizontal wind (u, v) at the height corresponding to that range can be retrieved as long as data contaminated by ground returns are omitted. The average wind profile over the large mesoscale area sampled by the radar generally evolves slowly. Consequently, while the results of the VAD wind retrieval can be plotted for a single time, it is more useful to examine the time evolution of wind as in Fig. 5.21. In periods of precipitation, the information is similar to what a wind profiler would measure, though applying to a larger area. Radar-derived time series of wind profiles are most useful for time periods between 6 and 24 h: below 6 h, limited time evolution is expected given the spatial scale over which the wind profile is made, at least in weather situations for which such a product is meaningful; beyond 24 h, the information obtained loses most of its relevance for forecasting uses. Unfortunately, on many radar data processing systems, such displays are often limited to 1 h.

It is theoretically possible to retrieve more information from Doppler data over mesoscale areas than simply the average wind profile. In one of the end-of-chapter problems, it is shown that a second harmonic in $\sin 2\phi$ and $\cos 2\phi$ can be fit, as a result of which terms related to the mean divergence and deformation can be retrieved over the area considered. Other techniques attempt to retrieve additional parameters from the azimuthal variations of Doppler velocity data. In practice, unless there are very good reasons to do so, for example, when retrieving winds from a tropical cyclone, the more parameters are fitted through sometimes ill-behaved Doppler data, the greater is the chance that a believable but completely misleading result might be obtained.

5.5 Data contamination and ambiguities

In this chapter, we have considered until now only good Doppler data, "good data" being defined as accurate measurements of the radial velocity of the wind at the location it is being displayed. In practice, Doppler data may be contaminated either by returns from targets not moving with the wind or by erroneous measurements.

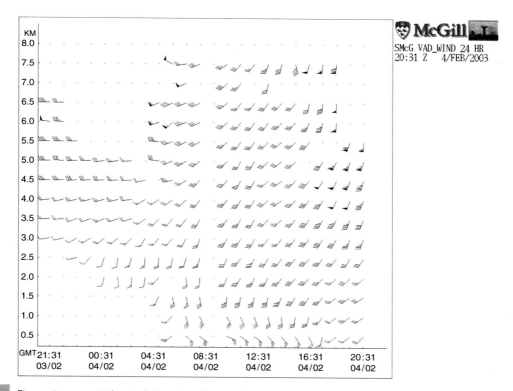

McGill
SMcG VAD_WIND 24 HR
20:31 Z   4/FEB/2003

Figure 5.21. Time series over a 24-h period of wind profiles retrieved by a scanning radar using the VAD technique during a midlatitude cyclone. Wind velocities are plotted using the standard convention, that is, a flag corresponds to 50 kt, a long barb to 10 kt, and a short barb to 5 kt.

Doppler measurements provide reliable information about the wind if and only if the targets observed move at the speed of the wind. This is true of refractive index discontinuities, hydrometeors, chaff, and to some extent summer daytime insects. Birds, migrating insects, and the ground or sea surface do not move at the speed of the wind and will therefore bias Doppler wind measurements. Ground targets bias velocities toward zero, and a good example is seen in the first image of Fig. 5.13 as well as in Fig. 5.22. If ground echoes are much stronger than weather echoes, the resulting velocity will be zero. If they are comparable in intensity, the result will be a weaker velocity than expected, something that may not be obvious to recognize. Some signal processing algorithms used in weather radars suppress zero-velocity targets, eliminating ground targets on both reflectivity and Doppler velocity imagery. Unfortunately, too often, they also reduce the reflectivity of weather with zero radial velocity and bias Doppler data away from zero for these echoes (Fig. 5.20). Birds, on the other hand, add their velocity to that of the wind. And if most birds within the radar coverage move more or less in the same direction, like during a migration, erroneous "winds" will be measured (Fig. 5.22).

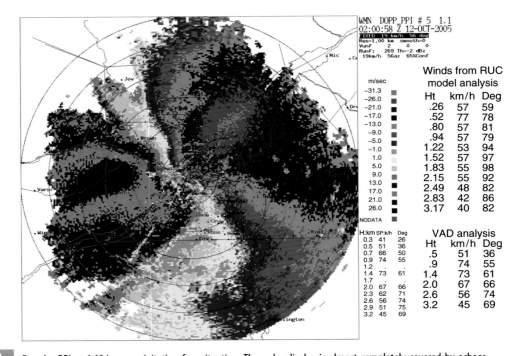

Figure 5.22. Doppler PPI at 1.1° in a precipitation-free situation. The radar display is almost completely covered by echoes from migrating birds, with a few patches of stationary echoes caused by the ground surface. A VAD analysis of the Doppler data, shown in the bottom right, suggests that the "wind" direction is primarily from 60° clockwise from north at 1.5 km. But true winds (indicated on the right as "winds from RUC model analysis") were primarily from the east (~90°); the difference between the two is due to the air velocity of migrating birds coming from the NNE (see Fig. 6.15).

On top of the biases caused by unwanted echoes, the introduction of Doppler processing has led to additional causes of data contamination that were generally not present when radars only measured reflectivity. These arose from the need to transmit pulses at a higher pulse repetition frequency (PRF) than in the reflectivity-only radar days.

To understand why, we need to revisit (5.2), the equation linking the change in the phase of targets and Doppler velocity, in the context of a radar firing pulses at a PRF f_r. The change in the phase of a target between successive pulses is

$$\Delta\varphi = -\frac{4\pi n}{\lambda}\frac{v_{\text{DOP}}}{f_r}, \tag{5.7}$$

where $\lambda = c/f$, the wavelength of the transmit pulse in vacuum. Phase change can only be measured unambiguously in the interval $[-\pi, \pi[$, a phase change of $-\pi$ between successive pulses being indistinguishable from a phase change of π. Velocities can hence only be measured over the interval $[-v_{\max}, v_{\max}[$ where

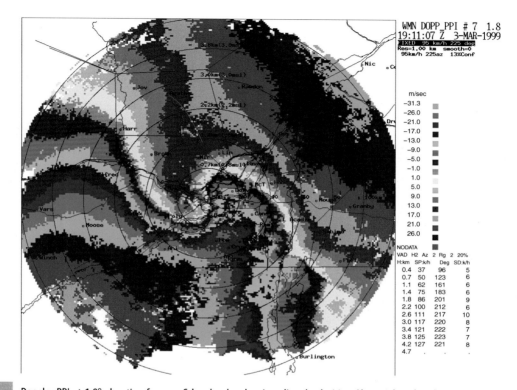

Figure 5.23. Doppler PPI at 1.8° elevation from an S-band radar showing aliased velocities. Above 2 km altitude, strong southwest winds exceed the 31.3 m/ s Nyquist velocity of this radar. As a result, velocities to the southwest and northeast of the display are aliased: in the southwest for example, strong approaching velocities that should be shown as negative velocities below −31.3 m/s (or in light blue) are instead shown as positive velocities (purple), and vice versa in the northeast.

$$v_{max} = \frac{\lambda}{4n} f_r. \qquad (5.8)$$

This critical velocity v_{max} is called the Nyquist velocity. For a C-band radar ($\lambda = 5.5$ cm) and a PRF of 1000 Hz, $v_{max} = 13.75$ m/s, a velocity often exceeded by upper-level winds. If uncorrected for, radial velocities exceeding v_{max} will be incorrectly reported, very very strong approaching velocities being indistinguishable from very strong receding velocities, and vice versa (e5.4). This phenomenon is known as velocity aliasing (Fig. 5.23).

According to (5.8), to increase v_{max}, two choices are at our disposal: increase λ, which generally implies the acquisition of a new and much larger radar, or increase f_r. But by increasing f_r, one limits the maximum range r_{max} at which the radar can observe echoes before the next pulse is fired:

$$r_{max} = \frac{c}{2nf_r}. \qquad (5.9)$$

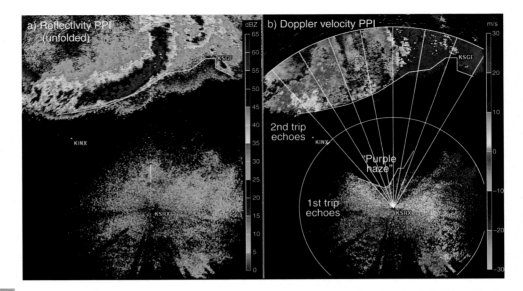

Figure 5.24. Example of the effects of range folding and velocity aliasing on WSR-88D data from Fort Smith (KSRX) on May 8, 2009, at 12:20 UTC. a) Reflectivity PPI collected with a low PRF (high r_{max}) and therefore without folding. b) Doppler PPI collected at a higher PRF (lower r_{max}). Circles are at r_{max} (125 km) and $2r_{max}$ from the radar, and the edge of the distant echo is outlined on both maps. Because these distant echoes are beyond r_{max}, the Doppler echoes should be folded, with the edge appearing where the near-range outline is drawn. In this case, a signal processing algorithm unfolded the strong distant echoes. Weak echoes that were detected on the reflectivity map but that were contaminated by the folding process have been marked in deep purple and are referred to in the United States as "purple haze." In parallel, despite the fast PRF of the Doppler scan, small areas of velocity aliasing can also be observed near the top of the Doppler PPI caused by the strong winds associated with this event (look for patches surrounded in white). See also supplement e05.4.

Echoes from ranges r beyond r_{max} will come back to the radar after the next pulse is fired and will appear as if they were coming from range $r - r_{max}$ (supplement e05.4). This phenomenon is known as the range folding of echoes.

Hence, if one fires pulses regularly, there is an inherent trade-off between v_{max} and r_{max}, or between the velocity aliasing and the range folding of echoes, referred to as the Doppler dilemma:

$$v_{max}r_{max} = \frac{c\lambda}{8n^2}. \tag{5.10}$$

Considering the previous numerical example for the C-band radar, we have $v_{max} = 13.75$ m/s and $r_{max} = 150$ km. Both are uncomfortably low since radar should be able to observe meteorological targets at ranges exceeding 300 km and with velocities greater than 40 m/s. And increasing one capability implies decreasing the other. This problem is less serious at the longer S-band wavelengths, but not uncommon (Figs. 5.23 and 5.24).

Over the years, techniques have been developed to recognize and try to correct for velocity aliasing and range folding. These include hardware techniques such as firing sequences of radar pulses at two or more different PRFs (Sirmans *et al.* 1976) and "phase coding" of transmit pulses (Frush *et al.* 2002), in parallel to software techniques to dealias velocities and flag doubtful data (like the WSR-88D "purple haze" on Doppler imagery). On many systems nowadays, the user can only see aliased velocities or range-folded echoes if they were not corrected by these algorithms or techniques. Even if significant fewer contaminated data points are left on radar images thanks to these algorithms, they have become as a result much harder to recognize.

Despite these contaminations, Doppler velocity data and derived products provide extremely useful information in a variety of weather situations. They are also not subject to calibration problems, a strength they have over reflectivity data. In a forecasting context, while reflectivity data are most useful for very short-term forecasting uses, Doppler data may offer the most useful source of information for longer term forecasting when used together with a weather forecasting model. And to help with calibration and especially data quality issues, one increasingly relies on dual-polarization data.

The added value of dual polarization

6.1 Why polarization matters

Operational weather radars have been undergoing another massive transformation. In addition to making reflectivity and Doppler velocity measurements, they are now collecting data at more than one polarization. This opens the possibility of measuring several dual-polarization quantities that can then be used in a variety of applications. Detailed treatment of the physics and the use of such quantities can be the subject of entire books, for example, Bringi and Chandrasekar (2001). In this chapter, only a brief introduction of some of the ideas behind multiple linear polarization measurements and of their applications is presented.

Virtually all man-made emissions of radio waves and microwaves are polarized. Most single-polarization weather radars transmit linearly and horizontally polarized EM waves (Fig. 6.1a). In the context of a wave traveling in the x direction in a (x, y, z) space, for such single-polarization radars, the electric field oscillates horizontally (in the y direction) while the magnetic field oscillates vertically (in the z direction). Hence, both the electric and magnetic fields oscillate perpendicular to the direction of propagation, assumed horizontal in this discussion. In contrast, most dual-polarization radars transmit waves at horizontal as well as at vertical polarizations, either alternatively or more often simultaneously.

Why is the polarization of the radar wave important? The interaction of the EM wave with hydrometeors and other targets occurs via the electric field of the wave, as the electric field in air induces a field in the hydrometeor. But the field in the hydrometeor will travel slower than that in air because of the higher refractive index. It is initially along the same axis as the wave in air but is then somewhat modified by the shape, size, and refractive index of the hydrometeor. The field inside the hydrometeor then interacts back with the field in air. This interaction has three results (Fig. 6.1b): (1) a partial reflection of the wave with a larger or smaller delay with respect to what would be expected for an idealized point target; (2) a partial attenuation of the transmit wave propagating forward because of loss of power from the reflection as well as from absorption; and (3) an additional delay of the forward propagating wave compared with what would have occurred in the absence of a target. If the shape of the target is nonspherical, and if the axis of oscillation of the electric field, or the polarization, is changed, the magnitude and properties of the wave–target interaction will also change. But in a nutshell, horizontally polarized echoes are more related to the

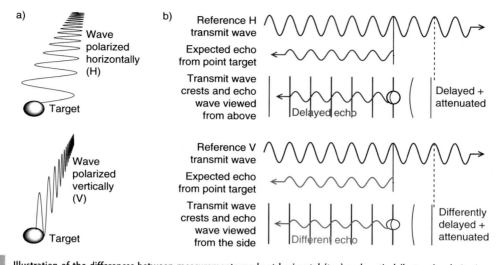

Figure 6.1. Illustration of the differences between measurements made at horizontal (top) and vertical (bottom) polarization. a) Illustration of the electric field oscillations at both polarizations. b) Comparison between the transmit wave (darker color) and its echo (lighter color) for an idealized point target and for a real target. Colored vertical lines represent the position of the crest of the transmit sine wave. As a result of the interactions between the transmit wave and the target, the transmit wave is delayed and attenuated when propagating forward past the target, while the echo may have a different phase than that of the ideal point target at the same range. Most importantly, these attenuations, delays, and echo strength are different at both polarizations; and their measurement provides new information on the shape and size of the target.

horizontal dimensions of targets while vertically polarized echoes are more related to the targets' vertical dimensions. By contrasting what is observed using one polarization compared with another, some information about the shape, and to some extent the size, of the target can be obtained.

All this would be irrelevant if targets were all spherical, or at least symmetric with respect to the axis of propagation of the wave. But individual targets are not symmetric: raindrops are roughly oblate or pancake shaped, and their flatness increases with size (Fig. 6.2); snowflakes have complex shapes and many orientations; and insects, birds, and ground targets are even less symmetric. These different targets therefore have different signatures in dual-polarization measurements that can be used to characterize what is being observed by radar. There lies the added value of dual-polarization radars.

6.2 Dual-polarization measurables

For single-polarization radars, Fig. 2.15 and the supplement e02.6 illustrated how the raw measurements being made are the amplitude of the signal returned by the targets and its

2.7 mm   3.45   5.3   5.8   7.35   8.0

Figure 6.2. Pictures of raindrops of different equivalent volume diameters. From Pruppacher and Beard (1970), © 1970 Royal Meteorological Society.

phase. What was illustrated was how, given a transmit pulse at one polarization, say horizontal (H), the radar measured the amplitude A_{HH} and phase φ_{HH} of the signal at that same polarization. When the returns are measured at the same polarization as that of the transmit wave, we refer to them as copolar signals. But if the radar can also receive signals at the orthogonal polarization, in this case vertical (V), then, given a horizontally polarized transmit pulse, it could measure the cross-polar amplitude A_{HV} and phase φ_{HV} of the signal. If in addition the radar can also transmit at vertical polarization, then the characteristics of the copolar signal A_{VV} and φ_{VV} and of the cross-polar signal A_{VH} and φ_{VH} can also be obtained. Hence, while single-polarization radars only use one time series of amplitude and phase at each range to obtain all their measurements, dual-polarization radars can have up to four. Most dual-polarization quantities are derived contrasting the horizontally polarized copolar echoes (HH) with either the orthogonal copolar echoes (VV) or the cross-polar echoes (VH). Many differential properties can be measured, and only a subset of them is presented here.

6.2.1 Differential reflectivity

The first dual-polarization quantity is the ratio between the reflectivity Z_H (or the power P_{HH}) measured at horizontal polarization from a horizontally polarized transmit pulse and the reflectivity Z_V (or the power P_{VV}) measured at vertical polarization from a vertically polarized transmit pulse. This ratio Z_H/Z_V or P_{HH}/P_{VV} computed from reflectivity or power in linear units is referred to as the differential reflectivity Z_{dr}. It is generally expressed in dB, a Z_{dr} of 0 dB meaning that $Z_H = Z_V$. In the absence of attenuation, Z_{dr} is a function of both the mean axis ratio of targets and of the dielectric constant of the scatterers $\|K^2\|$, dielectrically dense hydrometeors such as liquid water having larger Z_{dr} values for a given axis ratio than dielectrically light particles such as snow (Fig. 6.3). Since most nonspherical targets tend to be oriented with their longest axis horizontal, Z_{dr} is rarely smaller than 0 dB for atmospheric targets, hail being a notable exception.

Figure 6.4 illustrates the typical values of Z_{dr} values observed for different targets. By examining the drop shapes shown in Fig. 6.2 and the expected values of Z_{dr} with hydrometeor shape in Fig. 6.3, it can easily be deduced that in rain, high Z_{dr} values are indicative of large drops while low Z_{dr} values imply the presence of small drops. This opens the possibility of using Z_{dr} in conjunction with Z for a better estimate of rainfall, Z_{dr} contributing to reduce the uncertainty associated with drop size distributions (Seliga and Bringi 1976). As rain rates increase, Z_{dr} generally increases. But if dry hail is

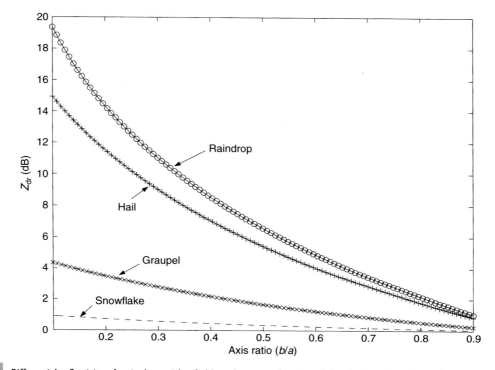

Differential reflectivity of a single particle of oblate shape as a function of the ratio between the vertical and horizontal dimensions of these particles. (Bringi and Chandrasekar 2001, © 2001 Cambridge University Press).

Chart of typical values of Z_{dr} for different types of targets. The colors correspond to the scale used on WSR-88D radar displays for Z_{dr}, while for each type of target, the range of colors represent the range of expected values. For hail, 2″ corresponds to 5 cm. Image courtesy of the NOAA Weather Decision Training Branch (WDTB).

present, Z_{dr} diminishes toward 0 dB or sometimes to negative dB values. This is partly because hail tumbles or occasionally falls with its largest dimension along the vertical. In high-reflectivity storms, a low Z_{dr} value is hence a reliable indicator of the presence of hail (Fig. 6.5). In snow, since Z_{dr} varies more slowly with the target axis ratio (Fig. 6.3), Z_{dr} is only high for elongated ice crystals. In parallel, insects look to the radar-like elongated water drops and generally have very high Z_{dr} (Fig. 6.5) that distinguish them nicely from drizzle of similar reflectivity.

6.2.2 Depolarization ratio

If the shape of a target is not symmetric with respect to the axis of propagation of the radar wave at one polarization, it will return some of the signal back to the radar at the orthogonal polarization. This phenomenon is referred to as depolarization, and it gives rise to the cross-polar signals mentioned previously. The magnitude of the ratio between the cross-polar signal (say, VH) and the copolar signal with the same polarization on receive (say, HH) is called the linear depolarization ratio, or LDR. Low LDR values, such as -35 dB (like all power ratios in radar meteorology, LDR is expressed in dB), arise from the returns of very symmetric targets such as small raindrops, while higher values rarely exceeding -15 dB in weather are caused by nonsymmetric targets such as hail. This is illustrated in Fig. 6.6, where rain and snow above the bright band appear symmetric and have very low LDR, while the melting layer, hardly visible on the reflectivity and differential reflectivity images in this example, can still be detected just above 2 km altitude as a line of higher LDR values. High LDR values are also observed associated with wet graupel as well as hail, and graupel is detected on top of the high Z_{dr} column at 68 km range. Finally, crystals aligned by an electric field will sometimes lead to a steady increase of LDR with range as seen at 60 km range and 8 km altitude in Fig. 6.6.

It should be noted here that not all radars measure all dual-polarization quantities. For example, many operational radars transmit both H and V waves together, as a result of which they receive at each polarization the sum of the copolar and cross-polar signals, for example, the HH and VH signals at horizontal polarization. Since the cross-polar signals are weak compared with the copolar signals for weather targets, it is assumed that only the copolar signals are being received. Such a radar design hence sacrifices LDR measurements for a simpler and more robust radar design that also leads to less noisy measurements of Z_{dr} and other quantities, simpler radar designs and less noisy data both being key requirements for operational radars that must scan the atmosphere rapidly.

6.2.3 Differential phase shift

A third quantity that can be measured is the differential phase shift between the phase of the copolar signal at horizontal polarization φ_{HH} and that at vertical polarization φ_{VV}. As illustrated in Fig. 6.1, this phase difference arises from two sources: a difference in the

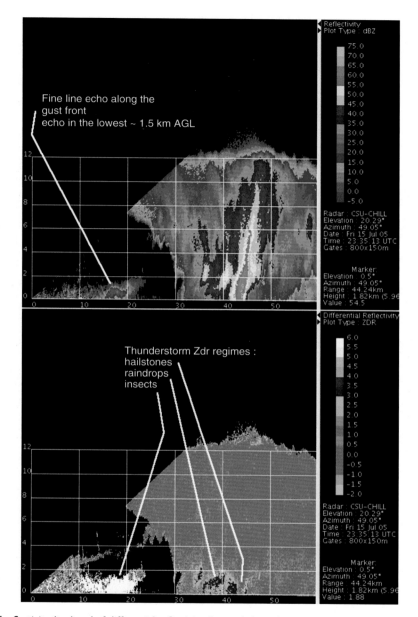

Figure 6.5. RHIs of reflectivity (top) and of differential reflectivity (bottom) through a convective cell in Colorado, USA, preceded by a gust front to the left. On the Z_{dr} image, one can clearly contrast the returns from hail (0 dB in a region of high reflectivity) and heavy rain (2–3 dB) as well as light rain (0 dB in a region of low reflectivity) and insects carried on the gust front (>5 dB). Image from the Colorado State University web site (2008), courtesy of Steve Rutledge.

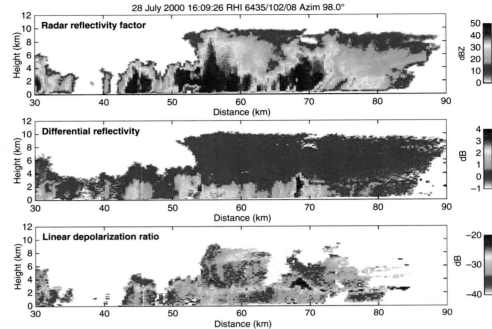

Figure 6.6 RHI of reflectivity (top), differential reflectivity (middle), and LDR (bottom) from Chilbolton, UK, through convective cells. Image courtesy of Robin Hogan.

delay introduced by the scattering of the transmit wave, known as the backscattering phase delay, and a difference in the forward propagation velocity of the two waves, known as the propagation phase delay. The differential phase shift ψ_{dp} between the phase at horizontal and vertical polarization at a particular range is

$$\psi_{dp} = \varphi_{HH} - \varphi_{VV} = \delta_{co} + \Phi_{dp} + \psi_o, \qquad (6.1)$$

or the sum of the differential backscattering phase delay δ_{co} of targets at that range, the two-way differential propagation phase Φ_{dp} that occurred between the radar and the target observed, and the phase difference between the two transmit waves at range zero ψ_o that is a function of hardware and software design specific to each radar. For Rayleigh and for spherical targets, $\delta_{co} = 0$; therefore, for most weather targets, the only term that varies with range is Φ_{dp}, as a result of which the symbol Φ_{dp} is often wrongly used for ψ_{dp}. Φ_{dp} generally increases monotonically but irregularly with range, as it reflects the characteristics of the entire path between the radar and the range of the observed targets (Fig. 6.7). Its rate of increase is a combined function of the number, size, and elongation of targets. A derived quantity is the one-way rate of increase of Φ_{dp} with range; it is called the specific differential propagation phase delay K_{dp} and is generally expressed in degrees per km. Except in heavy rain, K_{dp} is relatively small and hence difficult to

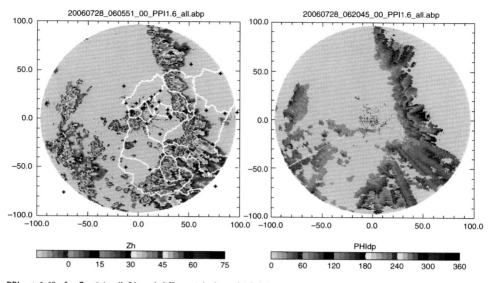

Figure 6.7. PPIs at 1.6° of reflectivity (left) and differential phase (right) in convective precipitation in Benin. On this X-band radar imagery, one can see differential phase increase with range in heavy precipitation cells in the east until rain completely attenuates the radar signal. Image courtesy of Marielle Gosset.

measure accurately given that it is a derivative (in range) of a difference (in polarization). Nevertheless, it has one crucial advantage over all reflectivity-based quantities: it is unaffected by attenuation.

K_{dp} is closely but not exactly related to rainfall rate. It is often quoted that K_{dp} is proportional to the number of drops of diameter D multiplied by $D^{4.24}$, instead of D^6 for reflectivity and approximately $D^{3.67}$ for rainfall; in reality, that number varies with wavelength and to some extent with diameter. Nevertheless, K_{dp} and Φ_{dp} have naturally been proposed to measure rain intensity: K_{dp} can be used directly (Sachidananda and Zrnić 1986) or in combination with Z_{dr} and/or Z (Ryzhkov *et al.* 2005) as will be shown below; Φ_{dp} measurements at far range may also be used as a constraint to determine Z–R relationships, calibrate the radar, or estimate path-integrated rainfall. At attenuating wavelengths, Φ_{dp}, a quantity unaffected by calibration or attenuation, can be used to help correct for signal attenuation. The interpretation of Φ_{dp} becomes more complex with targets other than rain. In ice-crystal regions, Φ_{dp} may increase or decrease with range depending on whether crystals are oriented horizontally or vertically (Fig. 6.8c). Insects and birds give varied and wavelength-dependent ψ_{dp} signatures depending on their size and species as well as viewing angle, only a subset of which have been properly documented at this time.

6.2.4 Copolar correlation coefficient

The last quantity measured by dual-polarization radars that is presented here is the correlation between the copolar HH and VV returns, called the copolar correlation

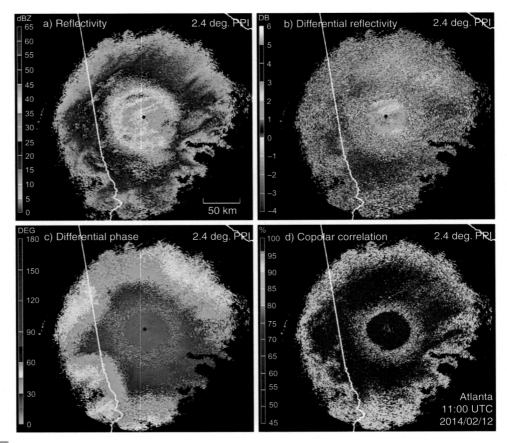

Figure 6.8.
PPIs at 2.4° of a) reflectivity (Z), b) differential reflectivity (Z_{dr}), c) differential phase shift (ψ_{dp}), and d) copolar correlation coefficient (ρ_{co}) in stratiform cold precipitation collected by the Atlanta WSR-88D S-band radar. Note how ρ_{co} is high in rain (within 50-km range) and in snow (60 km and beyond) but less so in the melting layer. Also note how differential phase stays more or less constant with range in this light rain but starts to increase in snow.

coefficient ρ_{co} (also referred to as ρ_{HV}). The range of values for ρ_{co} is between 0 (no correlation between the two signals) and 1 (perfect correlation). The electronic supplement e06.1 is useful to understand the complicated nature of this measurement. If targets have identical shapes, the patterns of constructive and destructive interference that give rise to the fluctuating radar signal will be identical at both polarizations, the only difference being a fractional change in the strength of the echo that is measured by Z_{dr}. If targets have varied shapes, the time sequences of returns at horizontal and vertical polarization become dissimilar: for example, a target configuration that would result in a perfect cancellation of the signal at one polarization will not do so at the other polarization. The time series of signals at horizontal and vertical polarization will not be perfectly correlated anymore as would be the case if the shape of targets were identical. And the greater the variability in shapes, or more precisely the variability in the Z_{dr} from

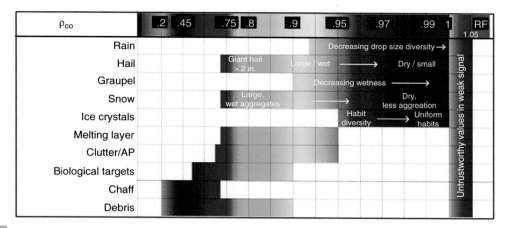

Figure 6.9. Chart of typical values of ρ_{co} for different types of targets. Image courtesy of the NOAA WDTB.

each target, the less the two signals will be correlated. And if targets are large enough to become non-Rayleigh, they may each have a different δ_{co}, further contributing to the loss of correlation. ρ_{co} is hence a good measurement of shape uniformity, sampling volumes with nearly uniform targets having very high ρ_{co}, while those that contain a variety of shapes, particularly for dielectrically dense targets such as melting snow and ground targets, have lower ρ_{co}.

Figure 6.9 shows the typical range of ρ_{co} values associated with different types of targets. Since ρ_{co} is a measure of shape diversity, it is a great indicator of complex situations. Rain and snow echoes have very high ρ_{co}, generally approaching 1; ρ_{co} values in melting snow are lower (roughly 0.80–0.90) and make the melting layer stand out (Fig. 6.8). Similarly, echoes from mixtures of hydrometeors associated with convection as well as those from large hail have the same range of ρ_{co} as in the melting layer; and if the sampling volume contains a significant number of large nonmeteorological targets such as birds or ground targets, ρ_{co} values will decrease considerably.

6.2.5 Other quantities

The imagination of researchers being the limit, many other quantities can and have been devised: reflectivity difference (as opposed to their ratio Z_{dr}), differential velocity, correlation coefficients involving cross-polar terms, etc. In addition, information on the texture or the "graininess" of some fields, such as the standard deviation of Z_{dr} or of ψ_{dp}, can be used to identify fields of echoes with different shapes such as birds or large clutter areas. A good illustration of the possible value of texture information is the small-scale variability of Z_{dr} in the melting layer in Fig. 6.8 compared with those in the snow above or in the rain below.

An interesting derived quantity that may become valuable is a proxy for LDR but derived using measurements available to radars that simultaneously transmit waves at both polarizations. Dubbed SDR (Melnikov and Matrosov 2013), and combining the

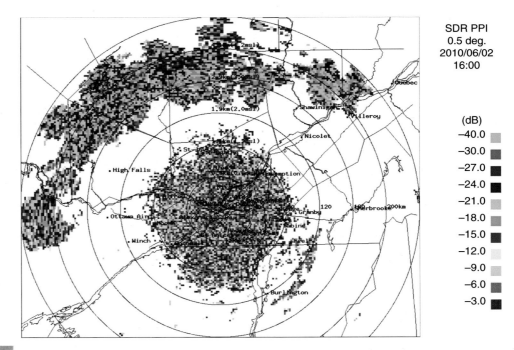

Figure 6.10. PPI of SDR showing echoes from ground clutter and insects (difficult to distinguish from one another using SDR alone) at close range and of precipitation to the north.

information from Z_{dr} and ρ_{co}, it was originally devised to estimate the axis ratio of ice crystals. But it is also proving to be a remarkable parameter to separate meteorological from nonmeteorological targets (Fig. 6.10).

6.3 Signatures and artifacts

In addition to reflectivity Z_H at horizontal polarization and Doppler measurements, we now have four additional measured fields at our disposal: the differential reflectivity Z_{dr}, the differential phase shift ψ_{dp}, the copolar cross-correlation ρ_{co}, and the depolarization ratio LDR, not to mention derived quantities such as Φ_{dp} and K_{dp}. They provide new information on the type of targets observed, targets whose spatial distribution is dictated among others by microphysical and dynamical processes. Therefore, these meteorological processes leave signatures on dual-polarization data that can be used to infer their presence.

6.3.1 Widespread weather signatures

A set of typical profiles of polarimetric measurements is shown in Fig. 6.11. In rain, the shape of raindrops is largely dictated by their size (Fig. 6.2), as a result of which the dual-polarization

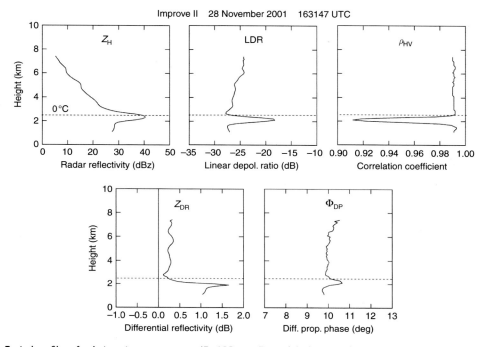

Figure 6.11. Typical profiles of polarimetric measurements (Z_H, LDR, ρ_{co}, Z_{dr}, and Φ_{dp}) in stratiform cold rain. The estimated 0°C level (2.47 km) for that event is shown by a horizontal line. Republished with permission of the AMS from Brandes and Ikeda (2004); permission conveyed through Copyright Clearance Center, Inc.

measurements follow predictable ranges. Z_{dr} is small for weak rainfall (between 0 and 0.5 dB at $Z_H = 20$ dBZ), increasing as the number and size of large drops increase to average about 2 dB at $Z_H = 50$ dBZ. LDR and ρ_{co} are, respectively, very low (<−25 dB) and very high (>0.98) because of the uniformity and near-sphericity of targets. As for Φ_{dp}, it only increases extremely slowly with range (Fig. 6.8). The melting layer appears clearly on dual-polarization data, with Z_{dr} and ρ_{co} having, respectively, a local maximum (a few dB) and minimum (of the order 0.9) in the bottom half of the bright band. These signatures peak where large melting snowflakes still have large-axis ratios and dielectric constants (high Z_{dr}), their large size and diversity of shapes leading to varied nonzero differential phase on scattering (δ_{co}), noisier Φ_{dp}, and lower ρ_{co}. Above the bright band, snow echoes show a wider variety of dual-polarization signatures depending on whether the echoes are dominated by roundish aggregates (low Z_{dr} and K_{dp}, often found at lower levels and warmer temperatures) or very elongated crystals (higher Z_{dr} and variable K_{dp} depending on the concentration of crystals, often found at higher levels and colder temperatures).

Because of its ability to determine the types of targets, especially wet snow, dual-polarization data are particularly helpful during periods of transitions between rain and

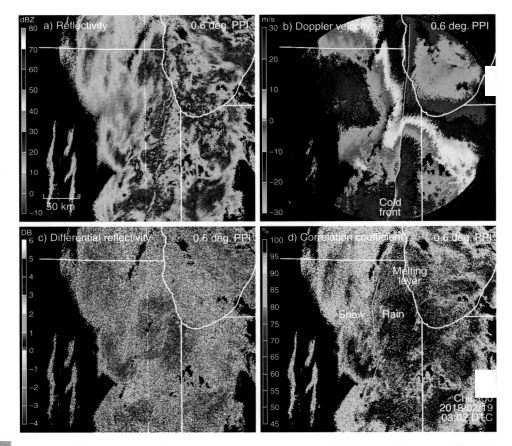

Figure 6.12. 0.6° PPIs of a) reflectivity at horizontal polarization, b) Doppler velocity, c) differential reflectivity, and d) copolar correlation coefficient during the passage of a winter cold front in the Chicago area. The reflectivity and especially the Doppler velocity imagery show that the cold front is east of the radar. However, Z_{dr} and ρ_{co} show that the melting layer ring, visible to the east, is truncated to the west where rain changes to snow. Also, occasionally, a line of Z_{dr} maximum and of ρ_{co} minimum can be seen at the location of the change itself; this line is only partly visible in this example, especially on Z_{dr}.

snow, particularly when those are driven by the passage of fronts. Figure 6.12 shows such an example. Precipitation had been from rain at the surface, and at the time of observation, the cold front had already passed over the radar and was 30 km to the east. But since temperatures behind the front dropped gradually, the transition from rain to snow was still 30 km to the west of the radar. This information on precipitation transition is extremely valuable, as some types of precipitation such as snow or freezing rain have greater impacts on public safety than others. It is, however, always critical to remember that all measurements are made at the level observed by the radar, and that further precipitation-type transitions may occur below the beam. Hence, especially when transitions are observed at greater ranges, the location where rain changes to snow aloft may not correspond to where it is observed at the surface.

6.3.2 Convective weather signatures

If considerable updrafts occur, supercooled liquid water can be lofted and maintained for long enough to coat falling graupel with liquid, raising both their reflectivity and especially their Z_{dr}. This gives rise to what are known as Z_{dr} columns above the expected 0°C level. One such column can be seen near kilometer 68 of Fig. 6.6, also associated with high LDR values and somewhat lower ρ_{co} (not shown). In this example, the column extends 1.5 km above the 0°C level, but they have been observed significantly higher in very severe convection, up to the −20°C level. Because these columns reflect the fact that considerable amounts of supercooled water are being lifted, they may be associated with regions where hail is growing and may fall later.

With the deployment of operational dual-polarization radars, a variety of potentially useful signatures are being explored (Kumjian 2013a). A particularly interesting one is caused by debris from tornadoes (Fig. 6.13). In essence, tornadoes lift debris with

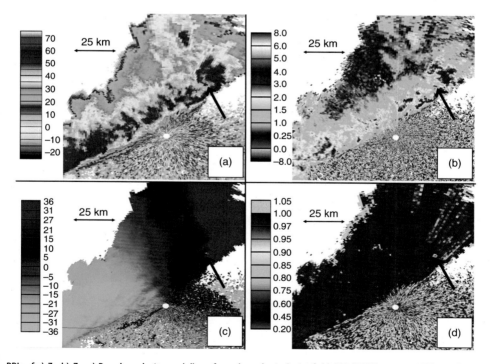

Figure 6.13. PPIs of a) Z_H, b) Z_{dr}, c) Doppler velocity, and d) ρ_{co} from the radar in Springfield, MO (KSGF), at 0605 UTC on February 29, 2012. Data show a tornadic debris signature, marked by the arrows, where a mesocyclone is being observed. From Kumjian (2013a), image courtesy of Matthew R. Kumjian.

complex shapes high enough to be observed by radar. Resulting echoes have very low ρ_{co} and sometimes near-zero Z_{dr} that are often collocated with mesocyclone signatures. The collocation of these two signatures is useful to confirm that the observed mesocyclone hides a smaller but much more destructive tornado.

6.3.3 Nonweather signatures

Nonweather echoes look very different on dual-polarization data than weather echoes. Not only are the values of some quantities such as ρ_{co} very different, but so is the texture of most dual-polarization fields. Figure 6.14 shows a moderately complicated example of dual-polarization imagery contrasting measurements from weather, insects, and emissions from the sun. If we only consider regions where the echoes are not very weak, say larger than 5 dBZ, we can see that both insects and the sun, as well as other nonmeteorological echoes not shown here, have low ρ_{co} compared with weather echoes. As also shown in Fig. 6.5, insects are easy to recognize by their very high Z_{dr} values caused by their elongated shapes. Sun emissions being uncorrelated with those of the radar, they have a measured ρ_{co} that is very close to zero while the differential phase shift is random.

Bird echoes are more complex to characterize simply, echo characteristics depending on the size and species of the birds being observed. Their complex shape also guarantees that their ρ_{co} is smaller than that of weather. Interestingly though, most dual-polarization quantities will change whether birds are observed from the front, from behind, or from the sides (Fig. 6.15). The details of how they change as a function of bird size, shape, and radar wavelength remain unknown at this time. But the richness of the information that can be obtained from birds and insects by radar is creating new uses for weather radars in ecology (Chilson *et al.* 2012).

6.3.4 Artifacts

With four new quantities being measured, there are new ways for measurements to be corrupted. A good discussion of the most common ones can be found in Kumjian (2013b). They can be grouped into two categories: measurement biases and propagation effects.

All radar estimates are made in the presence of microwave noise that comes from the atmosphere and from the radar itself. What limits the sensitivity of radars is when echoes become too weak to be detected amidst the noise. What this implies is that all radar estimates are affected by noise when the signal becomes weak. For some quantities such as Doppler velocity and differential phase shift, noise leads to less precise measurements. For others such as Z, Z_{dr}, and ρ_{co}, noise introduces biases, or systematic errors in estimates. This is often ignored for reflectivity as biases in very weak echoes have limited consequences. But biases in Z_{dr} and especially ρ_{co} complicate the interpretation of radar imagery. Noise has a near-constant Z_{dr} (typically around 0 dB) and near-zero ρ_{co}. Hence, a

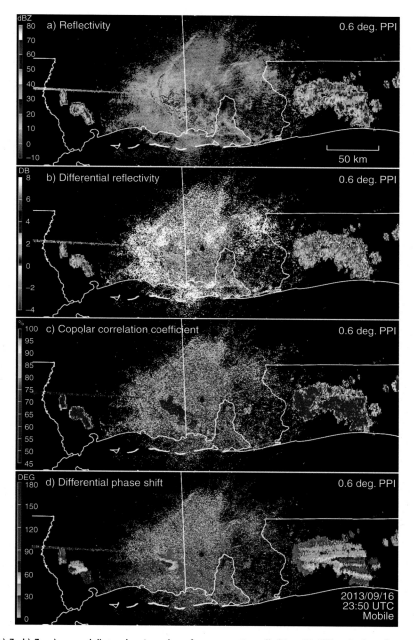

Figure 6.14. PPIs of a) Z_H, b) Z_{dr}, c) ρ_{co}, and d) ψ_{dp} showing echoes from convective cells ($Z_H > 20$ dBZ) and echoes from insects as well as emissions from the setting sun. Note the very high Z_{dr} and the lower ρ_{co} of insects, at least in regions where reflectivity is not very weak. Also note the smoothness of most dual-polarization fields in weather compared with those from insects.

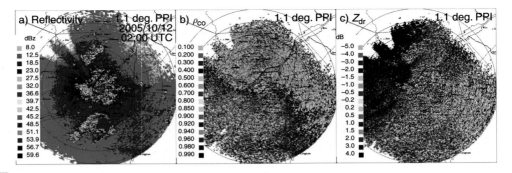

Figure 6.15. PPIs of a) horizontal reflectivity, b) copolar correlation coefficient, and c) differential reflectivity up to 120 km without any ground target filtering during a bird migration. The Doppler velocity image at this time was shown in Fig. 5.22. Lighter echoes (Z_H < 25 dBZ) are from migrating songbirds, stronger echoes being from ground targets. Observed ρ_{co} values are much weaker than for rainfall and vary with azimuth for birds depending on whether the radar observes them head-on (north-northeast) or from their back (south-southwest). A similar angle-dependent pattern is expected for Z_{dr} and for ψ_{dp} (not shown) and is partly observed on the eastern half. Remarkably though, the Z_{dr} for the western half are not symmetric with the eastern half, as if the birds' right side looks different to the radar than their left side. What is more likely though is that the bird species observed north and west of the radar are different enough from the species observed south and east to lead to a different angular Z_{dr} pattern.

few percent of noise in the data, undetectable in Z and even Z_{dr}, is sufficient to lower ρ_{co}, from values expected for precipitation (0.99) to values expected for melting snow or nonmeteorological echoes (<0.95). Given that radars are able to detect echoes even when noise somewhat exceeds signal, this can become a serious problem. Efforts must therefore be made to correct dual-polarization estimates for the effect of noise. On US radars, as of 2014, this correction is overdone, leading to ρ_{co} exceeding 1 when the signal is weak. This can be well observed at the edge of echo coverage on Fig. 6.8, where the pink color on ρ_{co} imagery corresponds to values larger than 1 in snow. In Fig. 6.14 though, that effect was more pernicious, weak insects unexpectedly having ρ_{co} greater than 1, complicating the interpretation of that image. The quality of the dual-polarization data at low signal varies from one radar model to another as it is a function of the signal processing algorithms used. A good test is to look at ρ_{co} imagery in precipitation on PPIs at a few degrees elevation like Fig. 6.8 and observe what happens as the signal becomes weaker. Any bias that is observed with signal strength will then have to be taken into account when interpreting other images.

Propagation effects are a bit more complex and have multiple causes and effects. Figure 6.16 shows an example of many of these associated with heavy precipitation. On this image, we can observe how differential phase shift ψ_{dp} (and Φ_{dp}) increases as range increases behind strong cells. This signature is expected based on what was discussed in Fig. 6.1. The first unwanted artifact is observed with Z_{dr}: in regions where ψ_{dp} values have increased considerably, streaks of negative Z_{dr} can be observed (in gray in Fig. 6.16). These are caused by attenuation, or more specifically differential attenuation:

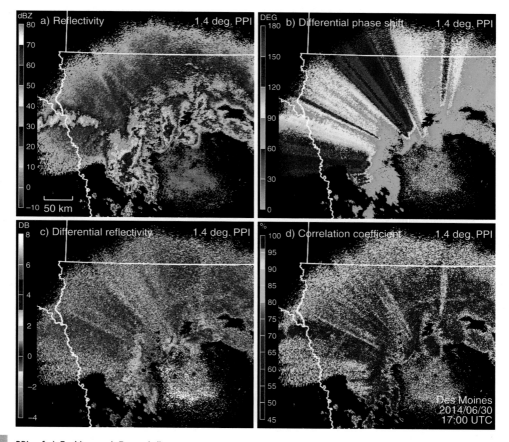

Figure 6.16. PPIs of a) Z_H, b) ψ_{dp}, c) Z_{dr}, and d) ρ_{co} in convection. In this example, Z_{dr} appears to drop below −1 dB behind strong cells (where ψ_{dp} is high), while anomalous streaks of lower ρ_{co} are generally observed where ψ_{dp} is varying rapidly in azimuth.

since horizontally polarized waves interact more with targets than vertically polarized waves in heavy precipitation, Z_H tends to be more attenuated than Z_V, leading to a systematic decrease of measured Z_{dr} behind heavy rain. The true values of Z_{dr} are not negative for this event, but they are measured negative because of differential attenuation. A second propagation effect arises when ψ_{dp} varies within the radar beam, either in azimuth or in elevation. Under these circumstances, the correlation between horizontal and vertical polarized returns is reduced, leading to streaks of reduced ρ_{co}. Note how ρ_{co} tends to be reduced where ψ_{dp} changes rapidly in azimuth, not necessarily where it is the largest. In both cases, behind severe weather, these effects are sufficient to change Z_{dr} and ρ_{co} from values expected in precipitation to values expected for other types of targets. To recognize these artifacts, it is always a good idea to be suspicious of every out-of-the-ordinary signature that is limited to a narrow range of azimuths and has the shape of a narrow piece of pie. All propagation effects such as attenuation and differential phase shift being stronger at shorter wavelengths than at longer wavelengths, expect these artifacts to be stronger at C- and X- band for similar weather conditions.

6.4 Applications

In the previous sections, we surveyed some of the properties measured by dual-polarization quantities. Even though the information presented was not organized by applications, a few should have sprung to mind: rainfall estimation, target identification, and data quality improvement. Historically, better rainfall estimation drove the interest in dual-polarization radars, but it may be the other two that will prove to be the most useful for operational use in the long run.

6.4.1 Rainfall estimation

With dual-polarization radars, we have at our disposal three quantities to help estimate rainfall: the traditional reflectivity Z, oversensitive to large drops; Z_{dr}, sensitive to the shape of medium to large drops; and K_{dp}, sensitive to a combination of the shape of drops and their number, and moreover immune to attenuation. A variety of methods for estimating rainfall have been proposed by combining one or more of these quantities. At the time of this writing, none has proven to be clearly superior to others under all conditions: techniques based on Z_{dr} are very sensitive to the calibration of differential reflectivity; K_{dp} is a noisy quantity; and all techniques rely on an assumed relationship between drop size and drop axis ratio that may be variable and is still being debated. Hybrid approaches where different relationships are used for various reflectivity or rainfall ranges seem to be gaining acceptance. For example, the US radar network uses different relationships depending on the types of targets measured and/or of rainfall intensity:

$$\text{Rain: } R(Z, Z_{dr}) = (0.0067)Z^{0.927}Z_{dr}^{-3.43}$$
$$\text{Rain-hail mixture: } R(K_{dp}) = 44|K_{dp}|^{0.822}\text{sign}(K_{dp}) \tag{6.2}$$
$$\text{Solid hydrometeors: } R(Z) = a(0.017)Z^{0.714},$$

where a can be 0.6 (wet snow), 0.8 (graupel), 1 (dry snow near the expected bright band level), or 2.8 (dry snow above the expected bright band level). Here, hydrometeor categories such as "rain" are determined by the target identification process to be described later. In experiments, the polarimetric relationships have been shown to provide better rainfall estimates than conventional Z–R relationships (Fig. 6.17). Best success is obtained if K_{dp} and Z_{dr} are measured accurately. In particular, noise or small biases in Z_{dr}, either due to a bad radar calibration or a difference in the attenuation between the horizontally and vertically polarized signals, can be extremely detrimental. Many other R–Z–Z_{dr} relationships have much higher negative exponents on Z_{dr} than (6.2); these are generally more accurate but more sensitive to noise and biases in Z_{dr}. Also note that at this time, radar-based

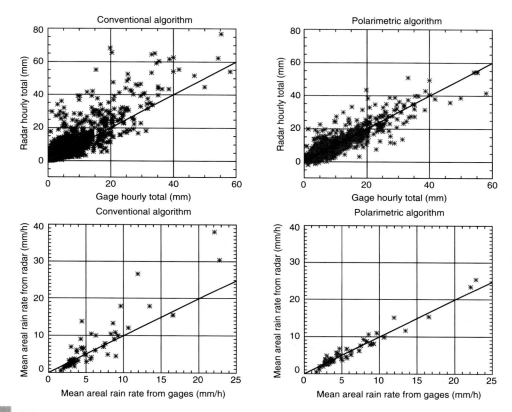

Figure 6.17. Radar-gauge comparisons using Z–R relationships (left) and polarimetric rainfall relations (right) for hourly data from individual rain gauges (top) and for all gauges combined within the radar coverage (bottom) in Oklahoma. Republished with permission of the AMS from Ryzhkov *et al.* (2005); permission conveyed through Copyright Clearance Center, Inc.

snowfall estimates do not use polarization information, as the shape distribution of snowflakes with size is ill defined.

Other approaches have been proposed to improve rainfall estimates with dual-polarization radars. One relies on Φ_{dp} measurements at far range to tune conventional Z–R relationships and correct for attenuation or radar calibration at the same time. The idea works as follows. After the radar pulse has passed through a sufficient amount of rain, a measurable Φ_{dp} can be easily observed. This Φ_{dp} at far range is a measure of the path-integrated rainfall and is relatively insensitive to both drop size distributions and attenuation. These properties can be used in at least three ways: (a) estimate the amount of attenuation from Φ_{dp} or K_{dp} and correct for it (Testud *et al.* 2000); (b) use the computed attenuation as a rainfall estimate (Ryzhkov et al. 2014); and, (c) use Z measured at each pixel to estimate an expected K_{dp} and, by integration, the expected Φ_{dp} along the entire path. The comparison of the latter with the Φ_{dp} actually measured allows an accurate calibration of the reflectivity measurements (Goddard 1994, Bellon and Fabry 2014, Fig. 6.18). Having corrected for any attenuation and calibration uncertainties, the

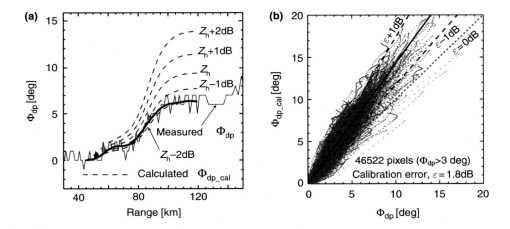

Figure 6.18. a) Comparison of raw measured Φ_{dp} (thin solid line) and smoothed measured Φ_{dp} (thick solid line) along an azimuth with Φ_{dp} calculated from Z_H (Φ_{dp_cal}, dashed lines) assuming different calibration biases in reflectivity measurements. b) Scatterplot of measured (x-axis) and calculated (y-axis) Φ_{dp} on the McGill S-band radar for one event. The solid line is a fit through the data used to estimate the calibration error ε. Two long dashed lines represent the possible range of the uncertainty of polarimetric radar calibration due to the variability of drop size distributions ($\varepsilon \pm 1$ dB). In this case, the data suggest that the Z_H data were on average 1.8 dB too high. From Lee and Zawadzki (2006), © Copyright 2006 with permission from Elsevier.

traditional Z–R or a polarimetric relationship can then be applied to obtain improved rainfall estimates.

6.4.2 Target identification and data quality

Weather radars are at their best when their information can be used quantitatively in an automated fashion. One of the great challenges is to ensure that the echoes observed truly originate from the expected targets, for example rain, and not from other sources such as hail, melting snow, or the ground. Contamination by unwanted targets is arguably the main cause of failure of precipitation and wind measurement algorithms. A reliable determination of the origin of an echo is hence a prerequisite to a successful quantitative use of radar data.

While changes in polarimetric quantities are relatively small in rain, they are much larger between different types of targets. In particular, nonweather echoes like insects (Fig. 6.3) have very different signatures that allow us to easily distinguish them: hydrometeors tend to have roundish-like shapes while insects, birds, and the ground surface have much more complex shapes. Hence, target identification using polarimetric quantities is a more robust task than to improve rainfall estimation using those same quantities.

That being said, there are always overlaps between the range of values observed for any given quantity for different targets (e.g., Figs. 6.5 and 6.9). The task of identifying targets therefore relies on how best to combine the information available from Z, v_{DOP}, Z_{dr}, ψ_{dp},

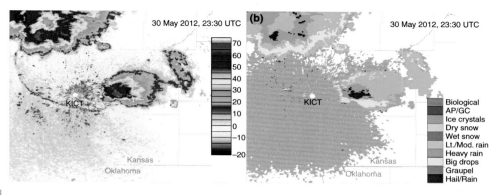

Figure 6.19. a) Horizontal reflectivity (Z_H) observed by the Wichita radar (KICT). b) Hydrometeor classification output based on that reflectivity field and dual-polarization measurements. In the classification, AP/GC stands for anomalous propagation and ground clutter (Kumjian 2013a, image courtesy of Matthew R. Kumjian).

K_{dp}, and ρ_{co}. A popular method for achieving this combination of information uses what is known as fuzzy logic (e.g., Park *et al.* 2009 and references therein; supplement e06.2). In a nutshell, for each variable or combination of variables, we test the likelihood that these observations fit with what would be expected for different types of targets. The strength of fuzzy logic lies in its ability to accept "maybe" as an answer by assigning a fitting function or a "membership function" from 0 to 1 for each test, 0 being "very unlikely" or "incompatible," 1 being "very likely" or "very compatible." Numbers in between 0 and 1 are used when the observed values are more or less likely for that type of target. For example, an observation of 5 dBZ reflectivity and 0.2 dB Z_{dr} would be very compatible with snow or drizzle (high score), reasonably compatible with light rain (medium score, because reflectivity would be a bit low), possible but unlikely with insects (low score, because the observed differential reflectivity would be unusally low for insects), and incompatible with hail (0 score). A more rigorous but more time-consuming approach is to compute the probability that each observation belongs to each type of target, and use these probabilities for the score. The process of transforming dual-polarization observations into test results from 0 to 1 has been given the name of "fuzzification." These tests are done for all variables, and other information such as expected temperature or time of day can also be considered. Once these different tests using all variables have been made, their results are combined, or "aggregated," using different weights for each test depending on which are judged more important or are better able to separate the targets into different categories. The target category with the highest score is generally deemed the winner, and the echo from that range cell is classified in that target category. Approximately ten types of targets such as drizzle, rain, hail, ground clutter, and nonmeteorological airborne scatterers can usually be distinguished by this process. An example of the result of an automated target classification is shown in Fig. 6.19. There exists many echo classification schemes that are similar in concept but use somewhat different "recipes" for classifying echoes. Some of the variability in approach can be explained because quantities such as ρ_{co} and K_{dp} vary with wavelength, signal processing, and how much the antenna scans while these quantities are being measured. Furthermore, some target categories such as "light" vs. "moderate" rain have subjective boundaries.

In addition to providing target identification, classification algorithms are used to filter the input to other algorithms. For example, precipitation rate calculation must be adapted to the type of target observed, and not made for insects or ground clutter. Echoes from unwanted targets must also be identified and avoided to improve the quality and usefulness of derived products such as VAD and the K_{dp}-based calibration algorithm. In general, the output from a good classification algorithm can provide better results in identifying the unwanted targets than most product-specific approaches. And while it is possible to classify targets to some extent without dual-polarization information, target identification algorithms based on dual-polarization data are much more robust. The artifacts mentioned in Section 6.3.4 though can lead to a mistaken identity of targets.

With this introduction to dual-polarization radars, we have completed the first part of this book dealing with the basic principles of weather radar and of its imagery. We are now ready to apply this knowledge and use radar for weather diagnosis and short-term forecasting.

7 Convective storm surveillance

7.1 A forecast approach

By its nature, convection is a relatively fast process. Most thunderstorm cells last less than an hour. Powerful updrafts and downdrafts are generated, resulting in locally heavy precipitation and strong winds. Several tens of cells can be present at once within the coverage of a single radar. They are organized in a variety of patterns and interact rapidly and in complex ways with their neighbors (supplement e07.1). This occurs as a result of the generally poorly observed interactions between storm flows.

Due to its short temporal and small spatial scale, convective weather is difficult to monitor without the help of remote sensing. Once it was realized that radar could provide data within storms almost instantaneously, it quickly became the instrument of choice for the surveillance of convective weather. This ability is the main reason why radar networks have been installed in many countries.

Upon reflection though, radar detects storms properly only very late in their formation process: it does not detect the destabilization of the atmosphere; rarely will it observe a storm's triggering mechanism, except at close range in the case of colliding outflows; it cannot detect the initial updraft, and except at extremely close range, it does not observe the burgeoning cloud. Only when the resulting precipitation starts to form do we have a first clue that something is happening, and only when a storm is mature can we make a full assessment of the situation.

To add to the difficulty of the weather surveillance and forecasting task, radar provides a large amount of data. Every hour, hundreds of PPIs of sometimes hard-to-interpret fields such as Doppler velocity or reflectivity from many types of targets are collected. Sifting through such data is a lengthy process. Yet, the fast evolution of convective storms implies that the threat detection must be done rapidly. Even with the help of automatic severe weather detection procedures such as the mesocyclone detection or the weak echo region (WER) identification algorithms, the correct interpretation of radar data takes time and must therefore be achieved as efficiently as possible.

The consequences of these two limitations, namely the relatively late detection of storm echoes and the vast amounts of data, are of crucial importance for forecasting. As long as radar remains the primary tool for convective weather surveillance and forecasting, we are essentially limited to warn for existing phenomena, or for phenomena that are likely to happen within the next half hour by extrapolating existing observations. This severely limits warning times. It hence becomes essential to rapidly recognize the development of severe weather threats. To minimize the time to recognize a threat, anticipating it by recognizing its early signs is critical. That is done most efficiently after a careful study of the pre-storm atmospheric environment, from which the types of storms and threats likely to occur have been predetermined.

An efficient interpretation of radar data hence requires that the forecaster knows what to expect if the forecast is correct. At the same time, he should also be open to the possibility that the forecast could be incorrect and not try to force an interpretation where or when it does not apply. One way to proceed for best success is to do the following:

a. Prior to the event, gather information on stability, winds, large-scale forcing, etc., that is expected to shape the weather during the forecast period.
b. Imagine the resulting time evolution of storms one should observe given the information collected.
c. Build a mental picture of what such a time evolution of storms would look like on radar. Different storm types have different morphologies and are associated with different wind patterns. These will often translate into distinctive patterns of reflectivity and Doppler velocity.
d. Ensure during the event that what is being observed is compatible with the mental picture one has built prior to the event. If this is the case, look for signatures of threats expected for that type of weather (mesocyclones, wind gusts, hail, etc.).
e. Rapidly go back to step (b) if what is being observed is not compatible with the mental picture one has built prior to the event. This is one of the most uncomfortable situations a forecaster has to face, but as is often joked, it is the kind of circumstances when one's salary is earned.

Here again, this may look a priori like an inefficient process. Can't a forecaster in (e) simply recognize the type of weather on radar and rapidly readjust? Ultimately, perhaps. But quickly recognizing an unexpected pattern requires knowing which reflectivity and Doppler velocity patterns fit the forecast and which do not, and finding them on radar displays, all this in a high-stress situation. This is achieved only with experience, which unfortunately is usually acquired only after it was needed the most. Hence, in spite of its apparent inefficiency, the process described above is robust until one has gathered enough experience to jump steps.

Putting into practice this forecast process requires a basic knowledge of convective storms and their evolution. We thus begin this chapter with a brief review of thunderstorm formation, evolution, and organization, and then concentrate on what possible threats are indicated by the various signatures on radar displays.

7.2 Severe convection and its controls

Ingredients needed for severe thunderstorms to occur include

a. a high convective available potential energy (CAPE), a manifestation of a conditionally unstable atmosphere over a large fraction of the troposphere;

b. some initial resistance to that convection, or convective inhibition (CIN);

c. a steady supply of moisture. If it comes via a low-level jet, so much the better;

d. some upper-level support, for example, in the form of a short upper-level wave advecting vorticity;

e. wind shear, especially at low- and mid-levels; and,

f. low-level convergence on a boundary such as a front or a storm outflow in order to locally break the CIN.

The first three ingredients are of a thermodynamics nature and are not well observed by radar, except perhaps for the low-level jet when targets are present. Radar is therefore only one of many sources of information to be considered for understanding the pre-storm environment. The last three ingredients are of a dynamic nature, with wind shear and convergence being best observed by radar, but again only in the presence of targets. Factors influencing those include synoptic-scale processes, air mass characteristics, and topography. Both sets of ingredients shape not only storm severity and type on an event-by-event basis but also storm climatology (Figs. 7.1 and 7.2, supplement e07.2) and its diurnal cycle (Fig. 7.3; supplement e07.2). As a result, forecasters must be aware of the local peculiarities in each region.

Figure 7.4 illustrates how convective storm type is controlled by the combination of instability, environmental wind shear, and the initial lifting mechanism. Instability determines the potential for air to accelerate vertically, while wind shear shapes storms and influences their evolution. When environmental wind shear is limited, mostly isolated cells that do not interact in organized ways are observed (supplement e07.3) except perhaps when a larger scale phenomenon such as a sea breeze is present. As shear increases, specific organized interactions are promoted, and storm systems generally become larger and more complex as well as more persistent. They also assume different shapes on radar displays (Fig. 7.5). Finally, the lifting mechanism such as a cold front, a sea breeze, or smaller scale convergence patterns,

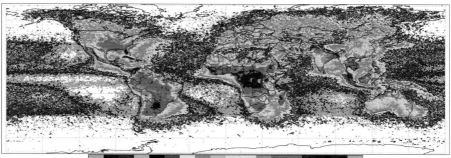

0.0 0.1 0.2 0.4 0.6 0.8 1.0 2.0 4.0 6.0 8.0 10.0 15.0 20.0 30.0 40.0 50.0 70.0 Flashes km-2 yr-1

Figure 7.1. Frequency of occurrence of lightning globally. This image provides an idea of the distribution of deep convection around the world in the absence of uniform global radar coverage. Global lightning image obtained from http://thunder.nsstc.nasa.gov/data/data_lis-otd-climatology.html, maintained by NASA EOSDIS Global Hydrology Resource Center (GHRC) DAAC, Huntsville, AL, 2013. Data for the image were provided by the NASA EOSDIS GHRC DAAC.

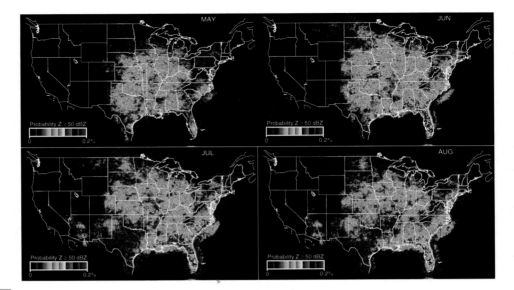

Figure 7.2. Probability derived from 15 years of US radar composites that an echo exceeds 50 dBZ in May (upper left), June (upper right), July (lower left), and August (lower right). Severe convection peaks in late spring in the Central Plains when the air is warm and humid enough near the surface and yet upper-level support is still important, later elsewhere where strong upper-level support is less critical. Thunderstorm occurrence is highest on the Gulf Coast and southern Atlantic Coast where sea breezes often play a major role in convection initiation, and lowest on the West Coast bathed by cold ocean water.

and the synoptic-scale environment associated with it also determine the location, shape, and extent of the convection. Even though all severe convective storms are capable of causing significant damages, each storm type is generally associated with a specific set of threats.

7.2.1 Evolution of simple convective cells (abridged)

An extreme simplification would be to say that convective storms often start with a cumulus cloud. Cumulus clouds form when "bubbles" (or parcels) of warmer air are lifted to the liquid condensation level (LCL), and then, if the atmosphere is conditionally unstable, to the level of free convection (LFC). After this point, parcels are free to rise beyond the equilibrium level (EL). Often, the EL is fairly low because of the presence of a capping inversion; the resulting cloud depth is hence small, and only a few drops may be generated (Fig. 5.5). But if CAPE is higher, the cumulus can shoot up rapidly (Fig. 5.1) and generate considerable precipitation. Wind shear then dictates the subsequent evolution of the storm cloud.

In the absence of significant vertical wind shear, the convective storm precipitation starts falling near the updraft feeding the main cell (Fig. 7.6). As rain partly evaporates during its fall, air underneath the thundercloud is cooled and sinks, forming a downdraft that diverges horizontally and nearly uniformly in all directions as it reaches the surface. As a result, the updraft and possibly its forcing mechanism are pushed aside, cutting the supply of moisture to the cloud that then starts to dissipate. The fact that the downdraft rapidly cuts off the storm's

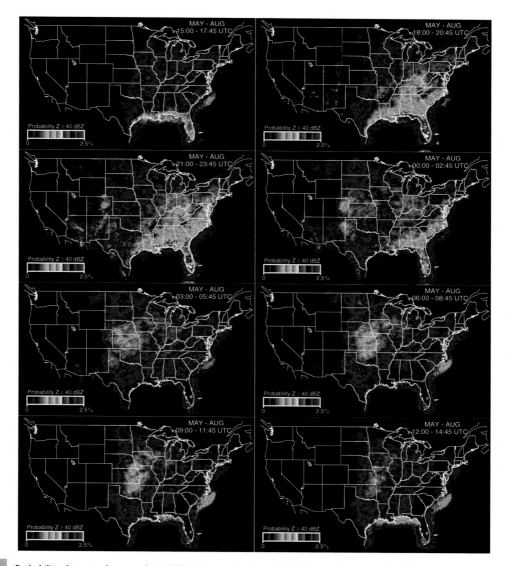

Figure 7.3. Probability that an echo exceeds 40 dBZ in the conterminous United States in summer for different times of the day starting in the morning (in central United States, 18:00 UTC corresponds to solar noon). Note at what time different areas experience peak likelihood of convection.

energy source considerably limits the lifetime and severity of such storms (supplement e07.3). Since both wind shear and winds at all heights are generally light, wind hazards tend to be limited, except if CAPE is large enough to cause the formation of a short-lived but powerful pulse thunderstorm that produces a microburst. In parallel, the light winds imply that a cell can remain over the same area for a relatively long period of time, possibly resulting in important rainfall accumulations locally. The main threat from single-cell storms hence comes from heavy precipitation, wind damage from microburst, and possibly short hail events.

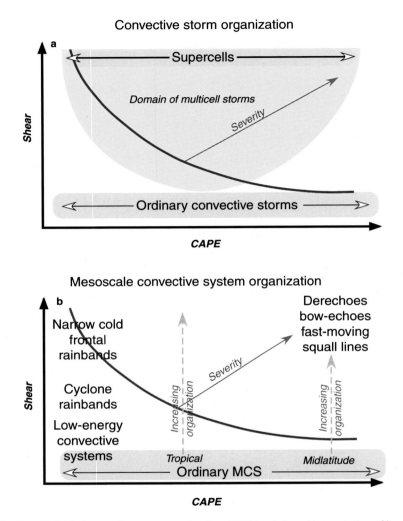

Figure 7.4. a) Organization of individual convective storms as a function of CAPE and shear; b) organization of lines and of systems of convective storms as a function of CAPE and shear. Republished with permission of the AMS from Jorgensen and Weckwerth (2003); permission conveyed through Copyright Clearance Center, Inc.

7.2.2 Vertical wind shear and storm organization

In the presence of stronger wind shear in the lower levels, storms evolve differently starting at the mature stage (Fig. 7.7). While the gust front moves away from the storm in all directions in a low-shear environment, it tends to remain more stationary upstream of a storm in a high-shear environment as a result of the balance between the opposing push from the storm outflow and that of the low-level environmental wind. The warmer environmental wind is then forced to rise by the cold outflow more or less at the same location for an extended period of

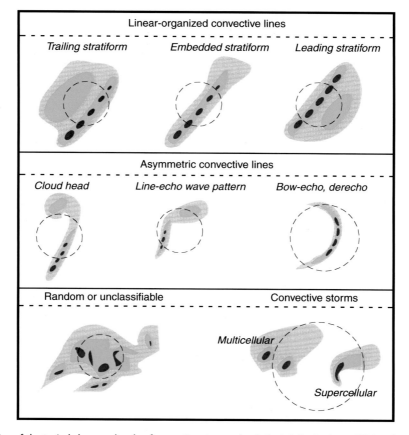

Figure 7.5. Illustration of the typical shape and scale of convective storms, the dashed circles having a 100-km diameter for reference. Republished with permission of the AMS from Jorgensen and Weckwerth (2003); permission conveyed through Copyright Clearance Center, Inc.

time, allowing the capping inversion to be locally broken and leading to the formation of a new cell upstream of the old one. This situation gives rise to multicell storms that are much longer lived than single-cell storms. A similar process also explains the longer life of many convective lines. Multicell storms often contain several cells in different stages of evolution. These storms move more rapidly because of the stronger winds, but the succession of cells can also bring considerable precipitation under their tracks. Wind damage becomes more likely as strong winds aloft can be directed by the downdraft toward the surface.

In the presence of strong curved vertical wind shear, a new storm organization may arise: the supercell (Fig. 7.8). In the supercell, the updraft enters on the right flank of the storm in the northern hemisphere (the front of the storm being the direction it is heading to) and feeds precipitation cores and their downdrafts both in the back and in the front of the storm. The horizontal expansion of the resulting low-level outflow is blocked by the inflow, reinforcing the updraft and gradually pushing it to the right in the northern hemisphere, or to the left in the southern hemisphere. In the end, what the curved shear allows is a clear separation of the

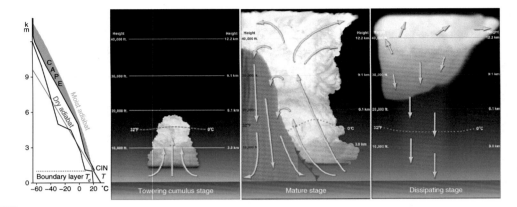

Figure 7.6. Idealized vertical section of the time evolution of a convective cell in a weak vertical wind shear environment during its growing stage, mature stage, and dissipation stage. Source: NOAA JetStream. Added on the left is a vertical profile of the environmental temperature (T) and of dew point temperature (T_d) associated with this event illustrating the vertical structure of the well-mixed boundary layer at the base of the troposphere, superposed with a zone of CIN offering some resistance to the formation of thunderstorms, itself superposed with a large conditionally unstable region with much CAPE.

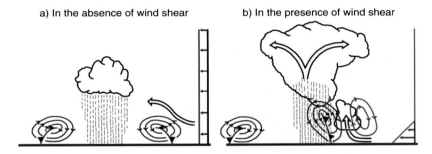

Figure 7.7. Cross-sectional diagram of a convective cell at the mature stage in a low-shear environment (a) and a high-shear environment (b). The edge of the cold pool, or gust front, is illustrated by cold front symbols, while the storm-relative winds are plotted on the right. In the absence of wind shear, the expanding cold pool hinders further storm development because the updraft is gradually pushed outward and cannot break the inversion. In the presence of shear, the expansion of the cold pool can be contained by the storm-relative winds at low levels, the latter becoming capable of reinforcing the updraft in one location so as to break the inversion and feed a new storm, or sometimes the same cell. Even though this illustration was designed for squall lines, it can be used to explain how shear causes all convective storms to last longer and be more powerful. Republished with permission of the AMS from Rotunno *et al.* (1988); permission conveyed through Copyright Clearance Center, Inc.; color version courtesy of George Bryan.

updrafts from the downdrafts in such a way that not only do they not interfere with each other, but they also actually provide mutual support. The result is a powerful storm that is self-sustained for long periods of time. Mature supercell echoes have a characteristic oval shape with a hook gradually developing on their trailing edge (Figs. 7.5 and 7.8), and often

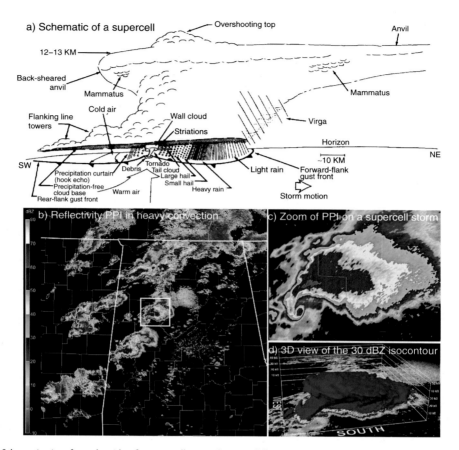

Figure 7.8. a) Schematic view from the side of a supercell storm (reprinted from Houze and Hobbs 1982, © Copyright 1982, with permission from Elsevier). b) Radar reflectivity PPI showing multiple supercell storms within the radar coverage. A supercell storm northwest of the radar is singled out by a rectangle. c) Zoom of the PPI over that supercell storm. Note the distinctive hook echo on the west side on which a strong mesocyclone signature associated with a tornado is also observed. d) Isocontour of the 30 dBZ surface in a 3-D perspective as seen from the south. On the vertical axis, 10 kft corresponds to 3 km. On the 3-D contour plot, note the weak echo region capped by an overhang east of the hook.

move in a direction that is more equatorward than less severe storms. Supercell storms can generate the whole spectrum of severe weather, from powerful downdrafts and strong surface winds to heavy precipitation and hail. They are however mostly renowned for their ability to spawn dangerous tornadoes (supplement e07.4).

7.2.3 Lifting by larger scale processes and convective lines

The storms discussed above are generally initiated over a small area by a local process that broke the capping inversion. However, lifting can also be forced by larger scale convergence patterns such as synoptic-scale fronts and drylines, as well as by mesoscale patterns such as

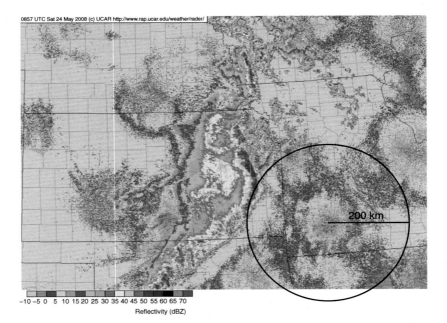

0857 UTC Sat 24 May 2008 (c) UCAR http://www.rap.ucar.edu/weather/rader/

200 km

−10 −5 0 5 10 15 20 25 30 35 40 45 50 55 60 65 70
Reflectivity (dBZ)

Figure 7.9. Radar composite image of a nighttime squall line in the Great Plains of the United States. The squall line at the center of the image is moving eastward, with heavy convection on its leading edge and lighter precipitation behind. Bluish disks east of the squall line are echoes from birds and insects at close range from different radars and taking advantage of the south winds ahead of the storm to migrate. For reference, the circle with a 200-km radius illustrates the typical reflectivity range of weather surveillance radars. Base image © 2008 UCAR, used with permission.

sea breeze fronts and topography. This often leads to the formation of long lines or bands of convective storms that often exceed in scale the coverage area of a single radar.

Often triggered by cold fronts, squall lines are long lines of thunderstorm that are often accompanied by lighter precipitation as they mature (Fig. 7.9). Squall lines can last 12 h, but heavy precipitation at any one location is usually of a much shorter duration unless the displacement of the squall line is mainly along its axis; for example, flooding would be expected if the squall line in Fig 7.9 were to be moving toward the northeast. When they move very quickly (>70 km/h) and are smaller (100–200 km in length), squall lines often take the shape of a bow (supplement e07.5), and powerful linear winds can be generated by the storms' outflow. Otherwise, they represent a danger similar to that of a typical thunderstorm, no more, no less.

7.2.4 The role of the boundary layer

As has been hinted at earlier, a key element for forming new convective cells is a mechanism to break the inversion located below the conditionally unstable region. Sometimes, larger scale processes such as fronts are sufficiently strong to force air upward so as to break that inversion, generally leading to lines of thunderstorms. At other times, "air

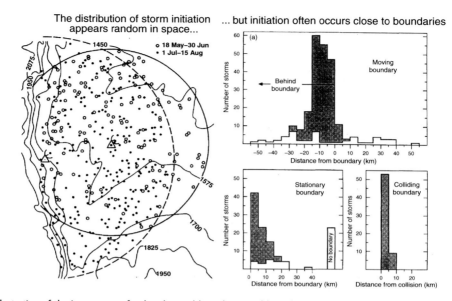

The distribution of storm initiation appears random in space... ... but initiation often occurs close to boundaries

Figure 7.10. Illustration of the importance of radar-observed boundaries and boundary layer circulations revealed by insect echoes for determining the location of convection initiation in the daytime: On the left, the spatial distribution of storm initiation appears random and independent of topography (contour lines represent altitude in meters), but the histograms on the right reveal that most storms initiate close to boundaries, dark shading outlining the events where the boundary is believed to have had a role in the initiation of the storm. Republished with permission of the AMS from Wilson and Schreiber (1986); permission conveyed through Copyright Clearance Center, Inc.

mass thunderstorms," as they used to be referred to, are generally formed thanks to updrafts that form below the capping inversion and that have enough kinetic energy and warmth to break this inversion. In the daytime, the region below the capping inversion is known as the boundary layer, a kilometer-thick layer above the surface. The boundary layer acts in two ways to help promote the updrafts required to break the inversion: first, it has circulations of its own, shaped by a combination of surface heating, temperature contrasts, mean wind speed, and topography. We can often see the result of these circulations thanks to insect echoes that tend to be stronger on updrafts (Fig. 4.15). Second, the boundary layer is the region where storm outflows propagate, and these cool outflows force warmer air to rise, in particular when they collide with fronts, dry lines, sea breezes, or other outflows (Fig. 7.10, supplement e07.6). In all cases, the key to the formation of new storms is that the updraft must persist under the same area of the capping inversion in order to first erode it, and then be able to supply warm and moist air to the growing thunderstorm. The sequence of events leading to the formation of a particular event can be extremely complex (supplement e07.7). Whereas at times it is possible to both observe these boundary layer processes and anticipate where and when new storms may form, that is usually not the case because either the radar data are not clear or time constraints prevent their proper interpretation. Consequently, we are currently mostly limited to warn for existing storms when using only radar data.

7.2.5 Threats

While some types of thunderstorms are more likely than others at spawning specific threats, for example, tornadoes, all thunderstorms can generate a wide spectrum of damaging weather. Severe storms may cause the following:

a. Damaging winds: Strong winds in thunderstorms occur either from the outflow from downbursts, generally causing linear or divergent winds, or from tornadoes causing rotating winds. Though tornadoes cause the strongest winds, both can produce significant damage. Tornado potential and strength depends on the amount of curved shear and CAPE; wind strength from downdrafts is dictated by the speed of movement of the storm, its intensity, and the relative dryness of the environment at low levels favoring evaporative cooling, the exact combination of which depends on local climatology.

b. Downbursts, and its danger to aviation: Airplanes flying through downbursts, especially during the landing stage, are in grave danger. Imagine an airplane flying through the base of the downburst. First, it encounters head winds, which increase both the air speed and the lift of the airplane, causing the pilot to reduce engine power and lower the aircraft. Then, it encounters the downdraft, immediately followed by tail winds, both pushing the aircraft strongly downward toward the surface.

c. Hail: Above a critical size (~2 cm), hailstones have enough energy to cause damage on crops, cars, and some structures, not to mention people. Airplane pilots also prefer to avoid hail for obvious reasons. Surface wind speed that adds horizontal momentum to hailstones is also an influential factor in hail damage potential.

d. Lightning: Lightning kills and injures people. While it cannot be directly detected with weather surveillance radars, certain echo characteristics like a reflectivity exceeding a certain threshold (~50 dBZ) above a certain height (6 or 7 km) are well known to be strongly associated with lightning activity. Lightning is also well observed by other dedicated systems.

e. Flash floods: Depending on the intensity of rainfall and its duration over any given point, heavy rainfall can cause damage through flooding. Floods generally occur when storms move very slowly and/or when multiple cells form or pass over the same point.

7.3 What to look for

7.3.1 Before the onset of storms

Before storms occur, we have to rely primarily on information provided by tools other than radar. Key questions to answer include the following: Are storms expected? Can they be severe and why? What types of storm organizations are likely to be observed given the amount of instability and shear in the environment, and given expected forcing mechanisms such as fronts (or lack of)? Once these questions have been answered, the most likely threats need to be determined. For example, are the conditions ideal for tornado formation and/or for damaging horizontal winds from downdrafts? Is hail or flooding a possibility? These are the same questions that must be answered prior to writing a proper storm watch for the day, if there is a need for it.

7.3.2 Threat identification during storms

When storm echoes appear, events precipitate, and the information provided by radar must be processed efficiently. First, are storms starting earlier or later than expected? If it is the case, what does it suggest about instability or capping inversion strengths, or the timing of triggers such as fronts? And how does that affect what type of weather and threats are expected and how we anticipate the latter sequence of events?

Then, one must be on the look for radar signatures associated with severe weather:

a. Very strong echoes ($\geq$55 dBZ) from precipitation at low levels: In a sense, this is obvious, but it is nevertheless worth repeating. Very strong echoes either mean very heavy rain ($Z_{dr} \geq 1.5$ dB), significant hail ($Z_{dr} \leq 0.5$ dB), or a mixture of both. If the storm is moving slowly, heavy rainfall warning thresholds may be exceeded (e.g., 50 mm in 1 h). The VIL product is ideal for differentiating cells with intense rainfall rates from those with only moderate potential. Wind gusts are also possible if the downdraft is strong enough. If the storm is moving rapidly, strong surface wind gusts become probable. Also remember that for the same hail size, more damage will be expected if horizontal winds are significant, hailstones gaining additional momentum from the horizontal winds. At attenuating wavelengths (C-band and shorter), rainfall rate estimates derived from a Z–K_{dp} combination can help detect heavy precipitation cores. At all traditional radar frequencies, a hail spike signature (Fig. 4.12) at mid-levels is also a great indicator of strong hail.

b. A very strong reflectivity core aloft ($\geq$50 dBZ at 4–5 km above the 0°C level, or 8-km altitude in midlatitudes; Fig. 7.11a): This is the first traditional criterion for severe storms from Lemon (1980). In many ways, it is a precursor signature to the previous situation: such a strong echo aloft can only mean that an extremely powerful updraft has lifted and condensed a considerable amount of precipitation aloft at cold enough temperatures that hail formation becomes likely, hail that will soon come down when the reflectivity core starts descending. It is also capable to bring with it a strong downdraft when it descends, possibly resulting in localized heavy winds at the surface. Here too, a high-level CAPPI or a VIL is helpful to quickly identify cells that potentially meet this criterion amidst a complex field of thunderstorms. Such a signature is more common in low-shear situation and for slower moving cells, but should be taken seriously in all situations.

c. Strong ($\geq$45 dBZ) echoes at mid or high levels (5–12 km altitude), overlapping much weaker echoes below from the same storm: More common in moderate- or high-shear situations, this is another traditional signature of a storm with a particularly strong updraft. In higher shear situations, it is not unusual for storms to become slanted. In the context of a strong near-vertical updraft, the base of the updraft would be precipitation free as no water

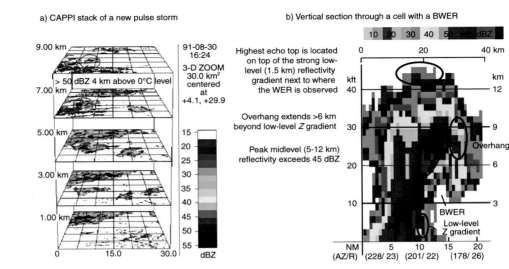

a) CAPPI stack of a new pulse storm

b) Vertical section through a cell with a BWER

Figure 7.11. Illustration of two criteria used to detect severe weather based on reflectivity patterns alone. a) Stack of reflectivity CAPPIs at different heights (1–9 km altitude) showing a growing pulse thunderstorm (outlined by gray ellipses) with a peak reflectivity exceeding 50 dBZ. In this case, the storm is just burgeoning and no precipitation has yet reached the surface. b) Vertical section of reflectivity through a particularly severe thunderstorm. Annotated are three criteria that must be met to declare such a storm severe; this particular storm vastly exceeded them all. This vertical section also illustrates nicely the WER underneath the precipitation overhang aloft. When the overhang descends beyond the WER, that weak echo is said to be bounded (hence "BWER"); this is a mark of violent storms.

has yet been condensed, but its top would have strong echoes; at the same time, the precipitation falling near the surface would have been condensed previously when the updraft was a bit away. The result is that the echo from such storms is slanted, showing what is known as a weak echo region (WER) below where the updraft is currently found at low levels, and an echo overhang above (Figs. 7.8d and 7.11b). If the updraft is particularly strong, the echo overhang will appear to spill down on the other side of the updraft, making the WER bounded with echoes from all sides. This signature is referred to as a bounded WER, or BWER. Lemon (1980) codified three criteria that must all be met for such storms to be deemed severe (Fig. 7.11b): (1) a strong enough reflectivity core above the melting level whose peak exceeds 45 dBZ; (2) the mid-level echo overhang must extend at least 6 km beyond the location of the strongest reflectivity gradient at the low levels (1.5 km high); and (3) the highest echo top must be on top of that low-level reflectivity gradient or on the overhang side of it. Supercell storms will often meet these criteria. It is important to ensure that these criteria are not met for the wrong reasons, for example, by combining elements coming from two distinct storms, or by an overhang that is simply the result of the fact that reflectivities at high levels were measured several minutes after those at low levels.

d. Storms or, worse, lines of storms, moving very fast (e.g., 80 km/h or more): Downdrafts from such storms are often accompanied by damaging linear winds at the surface. These may be observed on Doppler if the radar scans low enough and if the measurement geometry is right.

e. Strong storms moving slowly, or a sequence of storms passing over the same locations: This is a good recipe for generating copious amounts of precipitation that may trigger flash floods. Different forecast areas will have different thresholds for broadcasting heavy rain warning, such as 50 mm in 1 h. An important factor to remember is that rain rate is a strong function of reflectivity: 50 and 55 dBZ cells look very much alike on radar displays, but differ in rain rate by a factor greater than 2. Past and forecasted rainfall accumulation products are helpful tools to detect such potentially threatening situations.

f. Divergence signatures close to the surface (downburst) and rotation signatures detected over several elevation angles (mesocyclone): The former is a current threat for aviation; the latter could become a tornado and a severe threat to life and property. Automatic algorithms are often helpful for identifying storms with these signatures, simplifying the task of searching for them. But since these algorithms are not foolproof, the decision as to declare whether or not these storms represent a threat is generally left to the forecaster.

g. Tornado signatures: Tornadoes are extremely powerful small rotation circulations. If they occur very close to the radar, they may look like small mesocyclones (supplement e07.4). Their wind speed often exceeds the Nyquist velocity of the radar and their small scale (usually just a few hundred meters) make that rotation signature very hard to resolve. What may be more easily observed is a few pixels with extremely high spectrum width in the middle of the parent mesocyclone, betraying the presence of a strong turbulent circulation. Some help can also come from dual-polarization data, as debris lifted by tornadoes has random shape and therefore displays a very unique low copolar correlation signature and anomalously low Z_{DR} values for the given reflectivity; but in most cases, the data interpretation can be difficult at best, as illustrated in Fig. 7.12.

To help with the time-consuming task of threat detection, radar processing systems often have a variety of detection algorithms. The three most common ones are the mesocyclone, the downburst, and the WER (or conversely the overhanging precipitation) detection algorithms. All these algorithms are based on a conceptual model of storm evolution leading to severe weather and are triggered when the observations match what is expected from the conceptual model. For example, the most severe and longer lived tornadoes generally form within preexisting mesocyclones; therefore, an early warning for tornadic conditions starts with the prompt detection of mesocyclones. Since mesocyclones have a peculiar "kidney signature" on Doppler data (Fig. 5.15), the mesocyclone algorithm is designed to search for such signatures of appropriate size and magnitude through the Doppler radar data. Examples of implementation of severe weather detection algorithms can be found in Johnson *et al.* (1998) and Stumpf *et al.* (1998). These algorithms are very useful to help focus our attention toward a subset of storms that are more likely to be threatening, but they are not foolproof. Not all mesocyclones will spawn tornadoes, and not all kidney signatures are mesocyclones. Therefore, once an algorithm has identified a signature compatible with severe weather, it remains necessary for the forecaster to determine whether this is a potentially dangerous situation or a false alarm based on how the storm as a whole is evolving, and whether the signature identified makes sense in the context of the weather observed. Only after this has been insured should issuing a warning be considered. To minimize the chances of false alarms, radar data detection algorithms

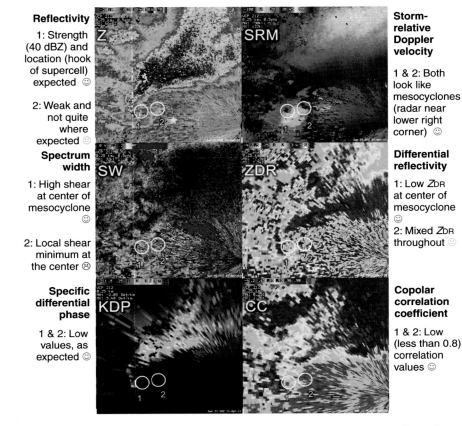

Reflectivity

1: Strength (40 dBZ) and location (hook of supercell) expected ☺

2: Weak and not quite where expected ☺

Spectrum width

1: High shear at center of mesocyclone ☺

2: Local shear minimum at the center ☹

Specific differential phase

1 & 2: Low values, as expected ☺

Storm-relative Doppler velocity

1 & 2: Both look like mesocyclones (radar near lower right corner) ☺

Differential reflectivity

1: Low Z_{DR} at center of mesocyclone ☺

2: Mixed Z_{DR} throughout ☺

Copolar correlation coefficient

1 & 2: Low (less than 0.8) correlation values ☺

Figure 7.12. PPI sections of six radar fields (reflectivity, storm-relative Doppler velocity, spectrum width, differential reflectivity, specific differential phase, and copolar correlation coefficient) in a supercell storm on which two potential tornado debris signatures with low ρ_{co} are being compared. While signature #1 meets all the criteria of a tornado debris signature, signature #2 has too many unexpected properties to be correct. Image is from WDTB (2013).

may be coupled to information from other sources such as model-derived information to help provide better warnings (Fig. 7.13).

7.3.3 Threat anticipation

Once the current threats are established, the regions that will shortly fall under their influence must be outlined. If possible, the occurrence of future threats based on current observations must be anticipated. In this respect, the following questions must be answered:

a. As a whole, have storm threats been increasing, staying constant, or diminishing? While it is difficult to accurately predict the evolution of any particular storm, it should be possible to determine whether, as a whole, the set of thunderstorms within radar

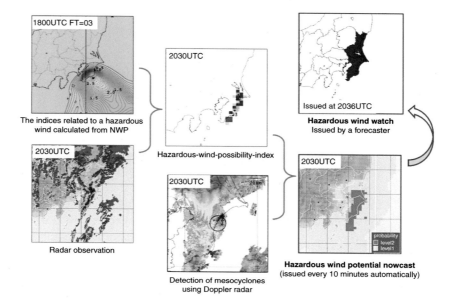

1800UTC FT=03

The indices related to a hazardous
wind calculated from NWP

2030UTC

Radar observation

2030UTC

Hazardous-wind-possibility-index

2030UTC

Detection of mesocyclones
using Doppler radar

Issued at 2036UTC

Hazardous wind watch
Issued by a forecaster

2030UTC

probability
level2
level1

Hazardous wind potential nowcast
(issued every 10 minutes automatically)

Figure 7.13. Example of a warning approach that combines radar data with other information. It is used by the Japan Meteorological Agency as guidance to warn for hazardous wind potential. From Joe *et al.* (2012), © Copyright 2012 InTech, licensed under Creative Commons 3.0 rules.

coverage is becoming more severe or not. Also, are storms generally entering or leaving the forecast area?

b. Are storms moving toward an area more or less favorable for storm development? Is instability or resistance stronger or weaker downstream? Climatology and familiarity with the local territory and larger scale forcing may be the only reliable source of information as radar is generally unable to provide an answer to this question.

c. Are storms heading toward areas that are more vulnerable? Some areas are more vulnerable than others, or can benefit from advanced warning more than others. Cities are a good example. Also, different areas are more or less vulnerable to different kinds of damage: cities and mountain creeks to flash flooding, forested areas to lightning, etc.

d. Which storms are or have the characteristics to become long-lived? To a first order, a better and a larger organization implies a longer lifetime. Individual supercell storms are also often long-lived.

e. Are new cells forming, or expected to form? This is probably one of the most difficult tasks to achieve. Satellite data may be of some help in the early stages of storm development before the images are filled by anvils and cirrus clouds. One of the rare occasions when storm formation may be successfully anticipated is when boundaries from clear air echoes collide (supplement e07.6), essentially guaranteeing that the warm air being squeezed by two colliding cool air masses be forced to rise, perhaps enough to break the capping inversion.

f. Can the combination of accumulated rainfall in the past and that of the immediate future exceed heavy rain thresholds?

To help with threat anticipation, radar processing systems again often have a variety of algorithms. Some track individual storms in the past and try to use that information to show their future position. Others may compute time sequences of forecasted precipitation. Forecasting for the immediate future is a topic of its own, and it will be further discussed in Chapter 10.

This overview was admittedly brief. There exists a considerable body of knowledge on the dynamics of many different convective systems and their associated radar patterns. That body of knowledge is continuously evolving as much research is devoted to the subject. Markowski and Richardson (2010) and Trapp (2013) propose solid introductions to mesoscale meteorology and storm structure, while Jorgensen and Weckwerth (2003) do the same with a research-oriented outlook. For operational applications grounded in theory, the convective weather sections of US training manuals such as the Convective Storm Structure and Evolution (Topic 7) Student Guide of the Distance Learning Operations Course (WDTB 2014) are more appropriate (note that the exact name of and reference for that manual has changed and will likely continue to change with time). In parallel, MetEd (http://www.meted.ucar.edu/) offers several training modules, in particular one on "Radar Signatures for Severe Convective Weather" that shows many examples.

Monitoring widespread systems

8.1 Radar, widespread systems, and threat forecasting

8.1.1 Time to reflect

Because radar has revolutionized the way convective storms can be monitored, it is primarily thought of as an instrument to be used for observing smaller, more rapidly evolving systems (Fig. 8.1). An undesirable consequence of this mental association between convective weather and radar is that when larger, or slowly evolving widespread systems are on the forecast menu, we often stop looking at the radar data and primarily rely on other sources of information.

Admittedly, there are at least two good reasons for downplaying the role of radar in widespread systems: (1) there are many additional tools to analyze widespread systems, from satellite imagery to surface and upper-air measurements, in addition to model-based prognosis; and (2) individual radar coverage is small compared with most widespread systems (e.g., Figs. 1.4 and 8.1), making it inappropriate for a complete assessment of the weather situation.

However, since widespread systems can also be disruptive, any instrument that can help assess the possible weather threats should be used. Moreover, radars being networked, it is now possible to get radar information over wider scales and to easily access data from radars outside of one's forecast region. Finally, forecasters have a luxury that they do not have when forecasting convective systems: time. Instead of just reacting to a fast-evolving situation, they can afford to reflect a few minutes and consider the more subtle yet rich information provided by radar under these circumstances.

As a result, this extra time can allow us to do more than just warn for an existing threat. Radar is most useful in widespread systems when it provides information that shows important discrepancies between what is expected to be observed given the current forecast and what is actually happening, and then provides additional clues on how to modify the forecast. Radar hence becomes less of a primary detection tool as it was in convection, but instead assumes a key role to help adjust forecasts, particularly in the 0- to 12-h time frame. This role, and the fact that warning for widespread weather threats does not require the immediate reaction needed in convective situations, call for a more analytical approach for using radar to its fullest potential.

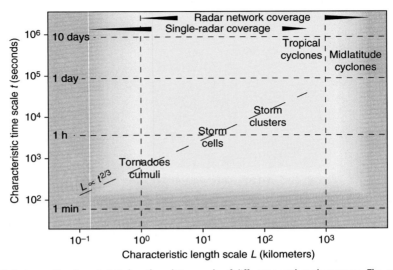

Figure 8.1. Relationship between the characteristic length and time scale of different weather phenomena. The gray-shaded area corresponds to space–time scales that are not well resolved by individual weather surveillance radar or radar networks.

Two very different types of widespread systems will be considered here: midlatitude and tropical cyclones. The main reasons they are grouped together in this discussion is that they both evolve relatively slowly, and that their forecasts from radar data are achieved in a similar manner. In practice, because tropical cyclones can have fast-evolving components such as embedded tornadoes and convective cells, a blend of a reactive and predictive forecasting approach is needed.

8.1.2 Defining weather information needs and threats

To help define a role for radar in widespread weather, it is important to recall what information is needed from a weather forecast. In many areas, two distinct types of users drive forecasting needs: aviation users with specific needs near airports and aloft, and a range of users from civil authorities to the public more concerned with surface weather.

Air travel is very much affected by weather. Around airports, aviation users require very specific wind, visibility, and precipitation information. For example, on takeoff, airplanes must be free of ice. During or after snow or freezing precipitation, they must be deiced. But the amount and type of deicing fluid to be used depends on the current and expected rate of snowfall or freezing rain during the next 15–30 min required for the airplane to get from the deicing station to the runway for takeoff. Aviation also benefits from wind and turbulence forecasts along their expected flight paths, and also from cloud and precipitation information around the airport of arrival. The same bad weather that delays the landing of airplanes can also cause icing on airplanes circling around airports if these fly through supercooled clouds or freezing precipitation. Therefore, to best help aviation, we need to observe and forecast wind and turbulence in the air as well as weather that may affect flight activities in the vicinity of airports.

Other users also need to know what can affect their activities or cause damages. Damages come primarily from precipitation and wind. Activities such as ground transportation can be impeded by weather when obstacles such as debris, water, snow, or ice lie on the roads.

Beyond what threshold can widespread events cause damages or major inconveniences? The answer depends on how different types and strengths of events affect the people and their activities in your forecast area. Different areas have different vulnerability thresholds to wind and precipitation, partly shaped by climatology and what type of weather events are expected to occur in any given area, as well as different adaptive capacity to deal with the situation. Also important is the extent with which forecast can positively affect that adaptive capacity.

Perhaps the best example to illustrate this point is to consider how precipitation affects activities and causes damages. We like to build close to water bodies, but not so close that our homes would flood every year; however, we often build in areas that are expected to flood less than once every 20 years. We also build infrastructure such as sewers to deal with "normal" rainfall, such as events that occur more frequently than once every few decades. Damage by flooding therefore generally starts when rainfall over the basin upstream exceeds what we call 20-year rainfalls, or the types of events one only statistically expects to observe at that location once every 20 years. Whether that threshold is 50 mm in 24 h in one location vs. 500 mm over the same period in another is largely irrelevant: the important factor is that the event is rare enough in that specific area for infrastructure to become at risk for water damage. And for basins beyond a certain size, it is primarily widespread events that are capable of causing flooding (Fig. 9.3). Then, one must consider area-specific issues: in mountainous regions, heavy precipitation can trigger other events such as mudslides, and those can cause damages of their own in addition to considerably impeding road and rail traffic. On the coast, heavy precipitation, inshore winds, and a high tide can combine to maximize damages. Finally, precipitation does not need to be that extreme to cause disruptions: it depends on the type of precipitation. Ten millimeters of rain will rarely affect anyone, but the same amount of water falling as freezing rain or snow, however, is another matter, especially for transportation. Here too, how problematic such events are also depends on whether these events are expected climatologically or not: the same snowstorm in Moscow or Rome will cause very different disruptions because one city has the infrastructure to deal with the snow and ice that the other one does not have. But the effectiveness of the deployment of such an infrastructure critically depends on the quality of the forecast as well as on how the forecast is interpreted and reacted upon, as sending snow removal crews too late or too early will diminish considerably their ability to deal with the weather.

In the end, different areas have different tolerances and vulnerabilities to different weather events. A proper forecast for disruptive weather must therefore start by establishing the thresholds beyond which wind, rain, or snow are likely to affect human activities and the built infrastructure. In addition, our ability to deal with such disruptive weather depends in the end on the accuracy of the forecast. And since much of the disruption caused by widespread systems is associated with winds and precipitation, radar with its reflectivity, Doppler, and dual-polarization measurements has a lot to offer.

8.2 What radar can contribute

Because widespread systems are large and evolve slowly, we generally have a good idea of the kind of weather and threats to expect before they start affecting our forecast area using model output or satellite imagery. The role of radar data is then to monitor whether the observed precipitation and wind patterns are compatible with the expected model forecast, and if not, to determine whether the discrepancies are significant enough to change the nature or the magnitude of the expected threats. The large-scale and slow evolution of these systems has two consequences. The first is that these systems are fairly well forecasted for the first 12 h where radar data can contribute pertinent information, and hence discrepancies between the forecast and the actual event are generally small. Finding those discrepancies requires some thorough detective work and is generally worthwhile only when small changes in weather patterns can lead to very different threat outcomes. The second consequence of the slow evolution of these systems is that any discrepancy is likely to extend over large areas. Therefore, radars located upstream can also be used to find these discrepancies, and thus gain a few hours to correct the local forecast: if the weather is not unfolding as expected 300 km away, it will likely continue along that trend further downstream; valuable information on how to change a local forecast may indeed be found in the neighboring region's radar.

Events for which small changes in expected evolution can lead to large changes in threats include the following:

a. Precipitation at temperatures near 0°C: Depending on the temperature profile aloft and the surface temperature, a variety of precipitation types, each with vastly different impacts, may occur at the surface: drizzle (warm rain process) or rain (cold rain process) that may be freezing or not at the surface, snow, ice pellets, and some rarer hydrometeors such as snow grains in the presence of convection. This is particularly true ahead of a warm front (Fig. 8.2): as a warm front approaches in cold winter climates, precipitation may first start as snow and then possibly change into ice pellets, freezing rain, and rain. The exact sequence, if any, will depend on front movement and on the intensity and depth of the cold air advection below the front and of the warm air advection above. Radar can be of an invaluable help in this complex situation: first, it can detect the appearances and disappearances of the bright band (Figs. 5.10 and 8.3). On occasions, a dual-polarization signature associated with freezing can be detected, though it often occurs at very low levels and thus may be difficult to observe (Kumjian et al. 2013). Radar can also reveal the height of the frontal surface, as it corresponds to the height where the winds are rapidly changing direction: a lower-than-expected frontal surface generally implies a thicker layer of warm air aloft and suggests a

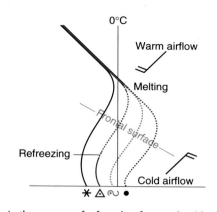

Figure 8.2. Vertical profiles of temperature in the presence of a frontal surface overhead leading to surface observations of snow (black curve), ice pellets (red), freezing rain (green), and rain (blue). Lines are solid for solid precipitation and dotted for liquid precipitation. Phenomena that lead to radar-observable signatures are annotated.

faster-than-expected transition away from snow in the future, the opposite being true if the frontal surface is higher. Finally, the wind speed and direction above and below the front can be measured and compared with expected values from model guidance. Stronger or more poleward winds above the front generally imply more precipitation and possibly a faster-than-expected transition away from snow; stronger or more equatorward winds below the front, that is, at and just above the surface, would reinforce the persistence of the cold air responsible for the refreezing of precipitation in that layer, delaying the arrival of nonfreezing rain. To make such assessments, it is advisable to examine higher angle PPIs, vertical cross sections and VAD-derived winds. However, rain and freezing rain cannot be differentiated by radar because the freezing occurs on contact with the surface; to distinguish them, temperature information at the surface or ground observations are needed.

b. Wintery precipitation: Once we know whether the precipitation to come or in progress is snow, ice pellets, or freezing rain, the next question is "how much?" Forecasts of amounts may prove inaccurate, either because the storm track is off by a crucial 50–100 km (which may represent less than 5–10% of the total distance traveled by a storm during the forecast period) or because of some unexpected local effects. In this case, comparisons of expected and observed precipitation patterns provide the most useful information (Fig. 8.4). Last but not least, the time of the expected onset of precipitation may need to be modified.

c. Possible flooding conditions: Tropical cyclones, and to a lesser extent strong midlatitude low-pressure systems, can generate large-scale floods (Fig. 8.5). If precipitation patterns move slowly toward downward sloping terrain, flooding can be exacerbated. Other surface conditions such as already saturated soil or snow on the ground can further enhance the impact caused by new rainfall. Though most widespread storms are generally well forecasted, small errors in the storm's trajectory and intensity can occur. These errors, or the unexpected evolution of some rainbands, may shift the location where peak rainfall

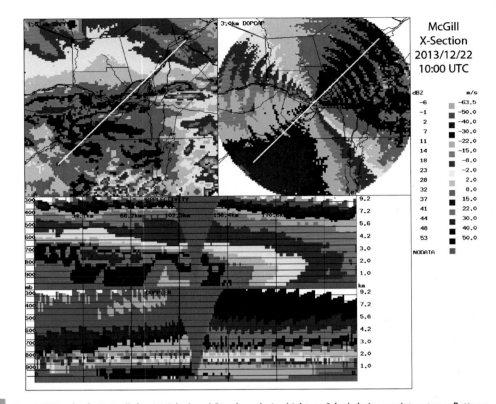

Figure 8.3. Top: CAPPIs of reflectivity (left, at 1.5 km) and Doppler velocity (right, at 3 km) during a winter storm. Bottom: Vertical cross section of reflectivity (above) and Doppler velocity (below) from the southwest (left side) to the northeast (right side). On the reflectivity cross section, the bright band is seen extending from the southwest to about 10 km northeast of the radar, ending on an east-to-west line of enhanced echoes on the CAPPI. In this case, precipitation at the surface is snow to the north of that line, and ice pellets and freezing rain to the south, surface temperatures remaining well below 0°C.

occurs: for example, note how the peak of precipitation after the first landfall of Irene in Fig. 8.5 is west of its expected location and of the storm track.

d. Damaging winds: We can generally predict correctly the intensity of surface winds, and where these winds may exceed critical thresholds. But the same small errors in storm track and intensification that may affect flooding forecasts will also affect damaging wind expectations.

e. Timing of wind shifts: Knowledge about rapid changes in wind direction is of great interest to aviation, in the air and especially on the ground. These are fairly well measured by radar: in cold fronts, where the wind shift first starts at the surface, their progression can be monitored on low-level PPIs (Fig. 5.14); in warm fronts, where the frontal surface is first seen aloft, a time series of VAD winds or an animation of a 2–4° PPI can often provide a good indication of when the front will reach the surface.

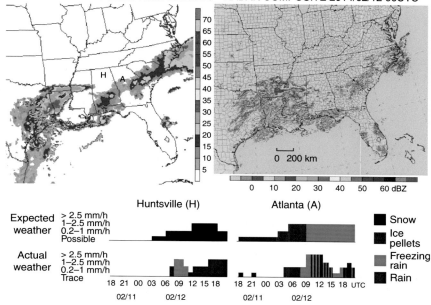

Figure 8.4. Top: Example of expected radar reflectivities from model guidance (left, courtesy of NWS) compared with observations (right, © 2014 UCAR, used with permission) for an intensifying winter storm in southern United States. The storm forming on the left side of both images is more displaced to the east compared with model guidance after just 6 h, suggesting that the model forecast may need to be updated. Bottom: Resulting time sequence of expected and observed precipitation in Huntsville and Atlanta, annotated as H and A on the forecast map above.

In addition to the list above, radars can help detect turbulence of various origins (Fig. 5.7) that may affect airplanes. In particular, warm frontal surfaces are a good medium to host shear-induced Kelvin–Helmholtz waves that can sometimes be seen on scanning radars (Fig. 8.6). Significant turbulence is also expected at the base of descending precipitation trails where snow is sublimating (Fig. 4.10). Also, embedded convection within large-scale precipitation structures such as cold fronts will also generate turbulence.

Finally, of particular interest to aviation are forecasts of icing. Icing caused by supercooled clouds (liquid clouds at temperatures below 0°C), freezing rain, and especially freezing drizzle can be severe threats, particularly to smaller airplanes. Such hydrometeors can freeze on contact with aircraft wings and control surfaces, affecting the airplane's ability to fly. There is no direct way of detecting whether clouds or precipitation is supercooled or not. But if the temperature is known, the problem reduces to being able to detect liquid clouds and precipitation. Supercooled clouds are nearly impossible to detect with the usual wavelengths of scanning radars because of

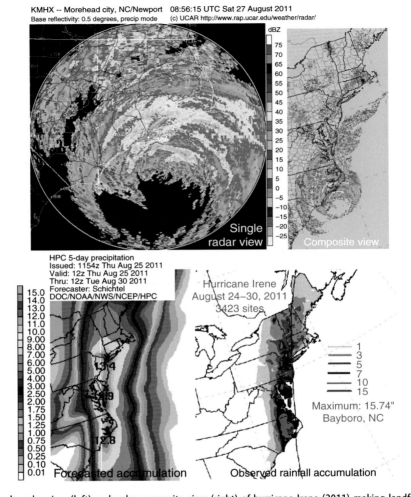

Figure 8.5. Top: Single radar view (left) and radar composite view (right) of hurricane Irene (2011) making landfall. Note how a single radar with a maximum range of 230 km can almost completely observe the most dangerous section of a hurricane (© 2011 UCAR, used with permission). Bottom: Forecasted (left) and observed (right) rainfall accumulation for hurricane Irene, together with its observed track (images courtesy of NOAA). Accumulations are in inches (1" = 25.4 mm). Note the area over which rainfalls exceeding 175 mm (7") were expected and observed, an amount comparable to the 24-h accumulation expected from a 100-year storm in the northern states.

their weak reflectivity. But their effect on snow may be detectable, particularly at vertical incidence: the same cloud that covers airplane wings with ice also coats snowflakes with ice, causing a slight increase in their terminal fall speed (Fig. 5.4). If these rimed snowflakes melt below, a weaker-than-usual bright band is observed.

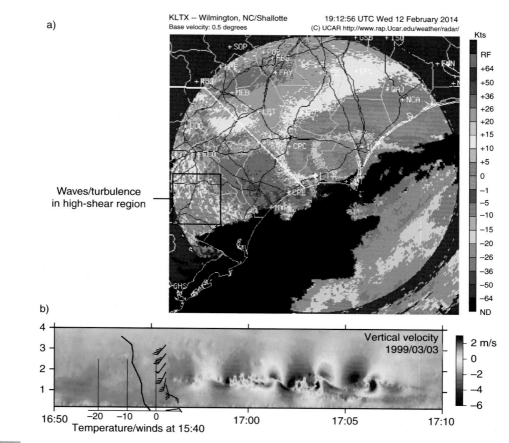

a)

Waves/turbulence
in high-shear region

b)

Figure 8.6. a) 0.5° PPI of Doppler velocity in a winter midlatitude cyclone. Winds shift rapidly from the northeast near the surface ahead of the warm front to the south around 2 km altitude and at the edge of radar coverage above the warm frontal surface. Near the western edge of the display, blue and yellow streaks of varying Doppler velocity caused by shear-induced turbulence can be observed. An aircraft reported moderate-to-severe turbulence in that area 20 min later (© 2014 UCAR, used with permission). b) Time–height section of vertical velocity in another event ahead of a warm front. Particularly strong Kelvin–Helmholtz waves can be observed around 1.5 km altitude where strong static stability and wind shear are observed. A closer examination reveals multiple wave trains of different scales at around 17:00. An airplane on final approach would experience successive vertical accelerations approaching ±1*g* in the worst locations.

Researchers are still looking for an unambiguous dual-polarization signature associated with that process. Freezing drizzle can also be difficult to distinguish because its radar signature is often similar to that of low-level light snow that can occur at similar temperatures. When embedded in snow, the presence of freezing drizzle can be inferred by analyzing Doppler spectra at vertical incidence (Fig. 4.4).

8.3 A case study of winter storm forecasting

The best way to illustrate many of the concepts described above is to apply them to a specific case study, given that they have not been well documented previously. Let us consider the meteorological situation on the evening of February 11, 2014, in the southeastern United States. A potentially crippling winter storm is in its developing stage and forecasts have been made based on model guidance. For the purpose of this exercise, we will focus on two cities that are expecting to see wintry precipitation arriving during the middle of the night (03:00–10:00 UTC in eastern United States), Huntsville and Atlanta. As the storm continues to approach, however, there is growing evidence that the forecast must be changed (Fig. 8.4). But how, and on what basis?

In Huntsville, the expected forecast was for precipitation to start around 6:00 UTC (Fig. 8.4). With the observed and modeled storms moving primarily from the west, the main area of mismatch between model forecasts and observations, 500 km to the southwest, should not affect the expected time of the start of precipitation at Huntsville. It had snowed early that morning from the previous storm that had then moved over the Atlantic, and the same was expected for the night and the day (Fig. 8.7), with surface temperatures hovering around 0°C. A hundred kilometers to the south, however, the elevated warm front ahead of the storm was expected to be sufficiently warm to melt the snow aloft; some numerical guidance even suggested that this could also be the case in Huntsville (Fig. 8.8): The North American Mesoscale (NAM) model forecasted temperatures at 850 hPa to be above 0°C and to remain so for several hours. There is hence disagreement among different forecasting tools as to the type of precipitation to be expected later that night.

Figure 8.9 shows radar data collected at 3:00 UTC, 3 h prior to the expected start of precipitation in Huntsville. The expected wind and temperature patterns were similar to those shown in Fig. 8.8 3 h later. A few discrepancies between expected and measured fields can be noted. First, the measured winds at 850 hPa (~8 m/s from the southeast on all three radars) are stronger than expected (~2.5–5 m/s), advecting more southerly air aloft over the region. The height of the 0°C level can be inferred from the height of the top of the ring of minimum copolar correlation coefficient around each radar. In the south and east, the 0°C level is significantly above 850 hPa, implying a warmer temperature than expected. In the west, however, the temperature at 850 hPa is colder than predicted, with the top of the bright band barely reaching 850 hPa immediately north of KGWX compared with the 2–3° C expected. That being said, melting is occurring around all three radars, even north of KGWX where snow was expected in the official forecast (Fig. 8.7). For all these reasons, the forecast of snow in Huntsville may be in jeopardy because the region of melting aloft is more north than expected, and melting is supported by stronger southerly winds than forecasted at 850 hPa.

In parallel, for Atlanta, precipitation was predicted to melt aloft and to refreeze close to the surface to form ice pellets. There is little sign of refreezing on the correlation

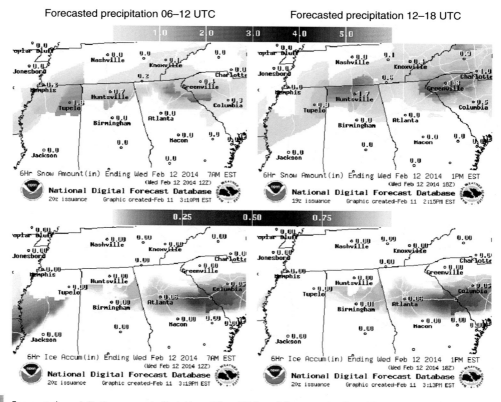

Forecasted precipitation 06–12 UTC   Forecasted precipitation 12–18 UTC

Forecasted precipitation amounts (in inches, 1″ = 25.4 mm) for snow (top) and freezing precipitation ("ice," bottom) for the end of the night (left) and the morning (right) of February 12, 2014. Images courtesy of NOAA.

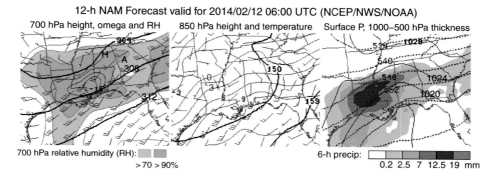

12-h NAM Forecast valid for 2014/02/12 06:00 UTC (NCEP/NWS/NOAA)
700 hPa height, omega and RH   850 hPa height and temperature   Surface P, 1000–500 hPa thickness

700 hPa relative humidity (RH):   6-h precip:
>70 >90%   0.2 2.5 7 12.5 19 mm

Model guidance for 6:00 UTC in the southern United States at the time when the storm is expected to reach Huntsville. Left: 700-hPa winds, height (in black contours, annotated in decameters), vertical velocity (omega, in brown), and relative humidity (colored). The location of Huntsville (H) and Atlanta (A) is indicated. Middle: 850-hPa winds, height (in black contours and in decameters), and temperature (colored lines, in °C). Right: Surface pressure (black contours, in hPa), 1000–500 hPa thickness (in colored contours and in decameters), and precipitation in the previous 6 h (colored field). Images courtesy of NOAA.

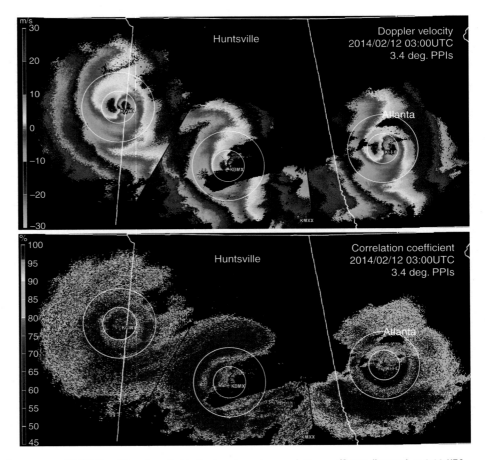

Figure 8.9. Composites of 3.4° PPIs of Doppler velocity (top) and copolar correlation coefficient (bottom) at 3:00 UTC from three radars south of Huntsville. Around each radar, two concentric rings indicate the altitude of the 850 hPa (1.3 km) and of the 700 hPa (3.05 km) levels.

coefficient images; but since the refreezing signature is difficult to spot under normal circumstances, this does not constitute a major discrepancy between forecasted weather and observations. Low-level winds are as expected, perhaps with a more northerly component, though it is not clear if they could advect the cold air required for ice pellets to form. But this argument is far from solid, and in reality there is little additional clue on radar to decide whether the ice pellets forecast is correct or not. Only the surface temperature and dew point temperature, both at or above 0°C, suggest that low-level temperatures may be too warm for ice pellets to be the dominant precipitation type at the surface.

While this example was made using horizontal maps, one can also compare model-derived soundings with wind profiles and the 0°C level obtained by radar(s). This can be

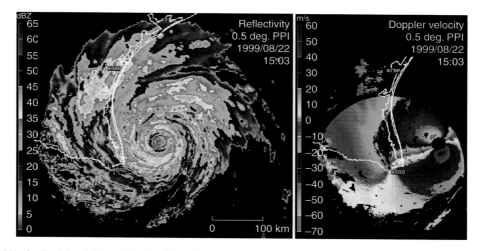

Figure 8.10. PPI of reflectivity (left) and dealiased Doppler velocity (right) from Hurricane Bret (1999) prior to landfall and measured by the KBRO radar in southern Texas. Be aware of the radar location to better compare both images.

done prior to the storm reaching the forecast area of interest using radars upstream as well as during the event if further changes in precipitation type are possible. Another example of the use of radar data in winter storms is shown in supplement e08.1.

8.4 What about tropical cyclones?

Figure 8.10 well illustrates the horizontal structure of tropical cyclones. In the center is the eye of the hurricane, a generally precipitation-free area. Immediately around is the hurricane eyewall, a band of convective precipitation near which the strongest winds can be found. The eyewall may be complete or partial, and for particularly strong tropical cyclones, two eyewalls may be observed. Beyond the eyewall are many rainbands that spiral around the tropical cyclone (see also supplement e08.2). On this storm, the spiral rainbands are evenly distributed all around the center of the storm, but it is not always the case. On the Doppler velocity image, at the center of rotation of the tropical cyclone, one can see what looks like a giant mesocyclone couplet. On this predominantly west-moving storm, approaching winds reach 65 m/s while receding winds do not attain 60 m/s, illustrating nicely that ground-relative winds in the Northern Hemisphere are strongest to the right of the storm if we consider the front of the cyclone to be its direction of movement. In vertical sections such as Fig. 8.11, the V-like structure of the eye is well outlined (be aware of the very different horizontal and vertical scales on the figure), surrounded by a strong convective eyewall extending to high altitude, itself surrounded by rainbands.

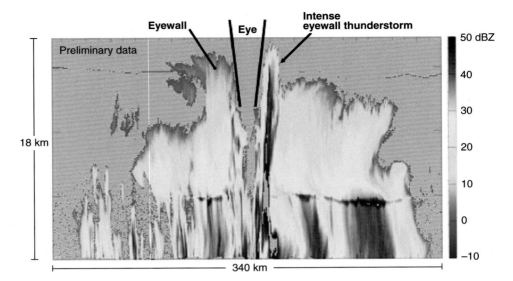

Eyewall   **Eye**   **Intense eyewall thunderstorm**

Preliminary data

18 km

340 km

50 dBZ
40
30
20
10
0
−10

Figure 8.11. Vertical section of reflectivity through a tropical cyclone taken by an airborne radar on board the NASA ER-2, a high-flying research aircraft. Image adapted from the original published on the NASA web site.

Like for midlatitude cyclones, the track and intensity of tropical cyclones tend to be reasonably well forecast. The role of radar is then to provide confirmation or updates to these forecasts and those of associated threats, wind damage, heavy precipitation, and storm surges. Changes in intensity can be monitored either using the Doppler winds or by examining at the height and intensity of the eyewall. Note that a near-perfect track and intensity forecast does not guarantee a perfect forecast on threats: for example, if the center of the storm is where it is expected, but rainbands are distributed in an unexpected asymmetric way, the position of the peak in precipitation may end up being shifted. This was the case for the hurricane shown in Fig. 8.5, whose first landfall was well predicted, but whose peak precipitation shifted west by about 100 km as the most intense rainbands, at the front of the storm at landfall, rotated to the left as the storm went inland. Locally enhanced floods can also occur in regions over which rainbands remain the longest. For example, on the left side of storms in the Northern Hemisphere, the spinning of rainbands and the forward movement of the storm can compensate each other in such a way as to maintain heavier rainfall over the same point.

It is not unusual for rainbands in tropical cyclones to become more convective, with echoes reaching intensities comparable to that of thunderstorms (Fig. 8.12). These convective cells can also generate small tornadoes along the front and right edges of the cyclone that add to the destruction. These small-scale circulations can be difficult to identify amidst the sometimes complex and very aliased velocities within tropical cyclones. Other more harmless circulations that can be observed include horizontal convective rolls that are also commonly seen in the planetary boundary layer (Fig. 4.15). Warning for these

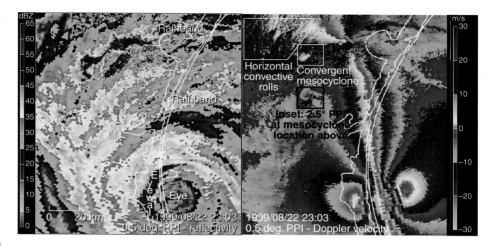

Figure 8.12. 0.5° PPI of reflectivity (left) and aliased Doppler velocity (right) from Hurricane Bret at landfall illustrating some of the smaller scale signatures observed in tropical cyclones. Among others, a mesocyclone is identified, and in a black-contoured inset is shown the 2.5° PPI imagery over that mesocyclone location. Contrast the true mesocyclone with the signature at the storm's center. And be careful of velocity aliasing when interpreting the Doppler image.

smaller scale phenomena is difficult, and we tend to react to them as we did with threats from convection, and not anticipate them as we do with other threats from widespread systems.

But the main threats from tropical cyclones remain the damage from storm surges and from flooding. The quality of storm surge forecasts is dependent on the quality of wind forecasts. Hence, ensuring that the wind pattern observed is similar to the one predicted is a necessary step when the time comes to make a forecast update. Large-scale flooding generally occurs over all regions visited by the tropical storm (Fig. 8.5), but additional smaller scale flooding also occurs in areas that experience a prolonged visit from a particularly convective rainband.

9

Radar estimation of precipitation

"It may be possible therefore to determine with useful accuracy the intensity of rainfall at a point quite distant (say 100 km) by the radar echo from that point." With that sentence, Marshall *et al.* (1947) launched the quest for the hydrological use of radar. And despite tremendous progress, this quest still continues. As you might have appreciated, radar is a superb tool for measuring the spatial patterns of reflectivity at the altitude where the measurement is taken, which, with some uncertainties, can be assumed to be a spatially distributed estimate of instantaneous precipitation rate. However, what is needed for hydrological purposes is a precise measurement of precipitation accumulation at the surface over a relevant time period. This is what rain gauges measure best: gauges are poor instantaneous rain rate instruments, but their measurement error diminishes rapidly with integration time. As will be explained in this chapter, this is not the case for radars. The art of radar hydrology hence consists in transforming a time sequence of instantaneous estimates of reflectivity and of dual-polarization parameters aloft into an unbiased estimate of precipitation accumulation at the surface. Properly integrating precipitation rates in time to obtain accurate accumulations requires a systematic fight against every source of error that can build up in time. These include systematic fractional differences for every measurement, or bias errors, such as a 15% underestimation due to a 1 dB radar calibration error. An equally problematic and less obvious challenge comes from errors that have long correlation times and distances, meaning that they are variable in time and/or space, but the time constant of their variation is slow compared with the accumulation time of interest. An example of such an error would be the one associated with extrapolating measurements made aloft at 1 km to the surface: climatologically, the difference between rainfall aloft and at the surface may be zero, but on any given day, there could be net evaporation or net low-level growth depending on low-level moisture availability. And we have learned over the years that the systematic fight against error sources is a challenging undertaking for which there are no shortcuts. But we keep trying, because radar provides an otherwise unavailable detailed picture of the spatial distribution of precipitation, while rain gauges provide point measurements (not necessarily bias free) that miss important details of the rainfall pattern needed to properly forecast floods.

9.1 Precipitation monitoring needs

An important first consideration before delving deeper into this subject is to determine the precipitation monitoring needs of a particular situation or of a particular user. Are quantitative precipitation estimates needed, or are qualitative or semiquantitative estimates sufficient? Radar is a naturally superb tool for obtaining semiquantitative estimates of rainfall: the dynamic range of precipitation rate covers about 4 orders of magnitude, that is, from 0.1 to 100 mm/h; even with a factor 2 uncertainty on rainfall (corresponding to a 5 dB error in reflectivity), radar can unambiguously identify $4/\log_{10}(2)$ or 13 clearly separable categories of precipitation intensities, enough to nicely map where different strengths of precipitation occur as a function of time. Therefore, if gross errors such as those due to bright band contamination are not an issue or avoided, a PPI of reflectivity that is not affected by attenuation can provide the needed adequate information. A large proportion of radar users fit in this category. The issues discussed in the rest of this chapter are more relevant to other users or situations that require greater accuracy.

Hydrological uses of weather radar data fit in two categories. First is the routine or systematic monitoring of precipitation for a variety of uses such as for climatological services, model validation, the management of rivers and sewers, and precision agriculture where irrigation and fertilizer use takes past and expected rainfall into account. Second is the early detection of flood for protection of life and property. Both of these uses have common requirements in terms of data accuracy, but the latter has some additional challenges because of the need to obtain estimates more rapidly, and sometimes not with all the information needed, in order to warn civil authorities as quickly as possible.

The key challenge is that relatively small fractional changes in estimated precipitation have large impacts: in many locations, a potentially catastrophic 100-year rainfall event may be only 50% greater than a still high but manageable 5-year rainfall event. This is partly due to the fact that the amount of water that stays at the surface is only a fraction of the total rainfall, the ground absorbing some rainwater in both cases; hence, a 50% difference in rainfall generally leads to a greater than 50% difference in surface water and outflow. At first glance, a 50% error on precipitation would appear to be a considerable amount. But in reality, we see that there are many sources of errors in radar precipitation estimates, and that their combined magnitude too often approaches or exceeds that number, especially if we consider that rainfall accumulations also span several orders of magnitude. The accurate estimation of precipitation is perhaps the most demanding task in radar meteorology in terms of its requirements on radar data quality.

Nevertheless, radar-derived hourly and daily accumulations are still useful because precipitation patterns can have considerable small-scale variability (Figs. 9.1 and 9.2). And with smaller basins, the likelihood of having other sources of information diminishes. Radar is hence particularly useful for small basins with short response times and where a sudden rise in river level can quickly overcome flow regulation infrastructure such as reservoirs and dams (Fig. 9.3). These basins first include the sewer systems of cities that act as a network of small rivers and where the costs of damages can climb rapidly. They also comprise mountainous watersheds that (a) experience localized heavy rains, (b) quickly channel the rain in gorges of raging water, and (c) often are not instrumented with stream or rain gauges.

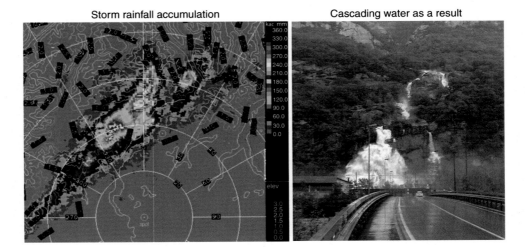

Storm rainfall accumulation   Cascading water as a result

Figure 9.1. Left: Zoom on an accumulation of rainfall over the duration of a convective event in the Alps close to the Italy–Switzerland border. Filled colors correspond to precipitation (in mm), while contours indicate terrain altitude (in km). Note the short distances over which rainfall varies from extreme to nonexistent compared with the range rings spaced 20 km apart (image courtesy of Robert Houze). Right: Photo of the resulting cascading water on the mountainside (photo courtesy of Matthias Steiner).

9.2 Sources of estimation errors

Errors in precipitation estimates from radar have been summarized in different ways by several authors, for example, Wilson and Brandes (1979), Joss and Waldvogel (1990), Chandrasekar *et al.* (2003), and Berne and Krajewski (2013). They can be organized in three categories:

a. Errors in radar measurements (reflectivity and dual-polarization data) aloft: These include (i) bias errors in measuring reflectivity and differential reflectivity due to a bad calibration, beam blockage by an obstacle, and attenuation; (ii) uncertainties in all parameters due to the fluctuating nature of radar echoes; and (iii) contamination by unwanted echoes and noise, as well as by the signal or data processing techniques designed to remove this contamination (Fig. 9.4).

b. Errors in estimating those quantities at ground level: Measurements aloft must next be transformed to estimates of these quantities near the surface. Precipitation can evolve considerably thanks to changes in the phase of hydrometeors (from ice to liquid water, for example) and to a variety of microphysical processes (low-level and orographic growth by accretion of cloud droplets, evaporation). Therefore, in order to achieve our goal, we first need to know what is measured aloft (relative proportions of snow, rain, hail, or bright band) and what processes are likely to affect

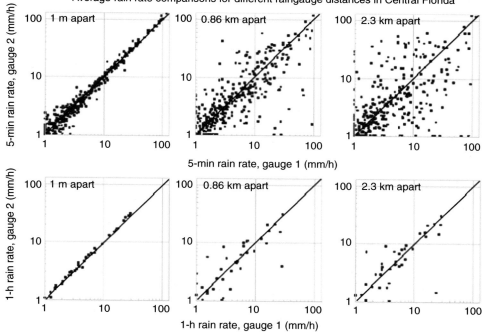

Figure 9.2. Comparison of the average rain rate computed over 5 min (top) and 1 h (bottom) measured by two identical rain gauges spaced 1 m apart (left), 0.86 km apart (middle), and 2.3 km apart (right). This illustrates both the variability of rainfall in space and the difficulty of measuring with a single gauge the precipitation seen by radar whose sampling volume can reach up to 1 km^3. Note the log–log scale of the plots. Data from Habib and Krajewski (2002); image courtesy of Witold Krajewski.

precipitation intensity with height for a particular event given storm dynamics and topography. Precipitation does take some time to fall down, an interval during which it can drift and sediment (Fig. 9.5), and the precipitation measured aloft at an instant is not the precipitation currently falling on the ground.

c. Errors in computing rainfall from the estimated radar quantities at the ground: Finally, with the estimates of radar quantities near the surface, one must derive a set of precipitation rate maps based on assumptions on how reflectivity, differential reflectivity, and specific differential phase are related to rainfall. These maps are then used to compute the final accumulation. Because of all the remaining errors and biases, gauge-based adjustments may then be applied, their accuracy depending on the representativeness of the gauge measurements of the precipitation field at the point of comparison.

There is an inevitable coupling in the errors from these three categories: for example, microphysical and dynamical processes dictating the evolution of precipitation affect both

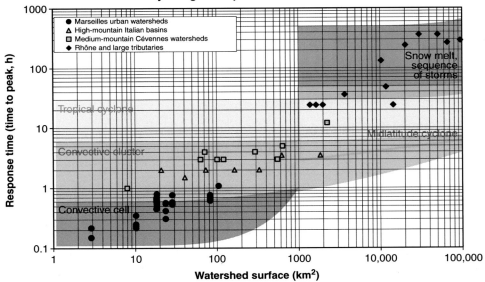

Figure 9.3. Response time, or time to peak flow, of basins as a function of their area (symbols) contrasted with the space–time domain influenced by different types of flooding processes (colored areas). The response time data plotted are for Mediterranean basins and are from Delrieu *et al.* (2014); other watersheds in less complex terrain may have somewhat slower response times. As the area of a watershed increases, so does its response time; consequently, the type of weather events that can flood that watershed changes. In parallel, the contribution of radar to hydrology is more important for basins in the lower-left part of the diagram, as other data sources (rain gauge and stream gauge) become more available when temporal and spatial averaging increases.

the particle size distribution (and hence the *Z–R* relationship) and the change of precipitation with height. Last but not least, an improper correction at an early step will adversely affect the ability of properly performing subsequent adjustments.

9.3 Radar-based accumulation generation

As was hinted above, the derivation of accurate precipitation estimates from radar is challenging. Considerable efforts have been deployed to provide solutions to most of these problems (e.g., Fulton *et al.* 1998, Germann *et al.* 2006, Tabary 2007, Harrison *et al.* 2012), but rare are the operational systems that successfully correct for all of the major errors. In the following discussion, we will limit ourselves to some of the most publicized solutions to key errors.

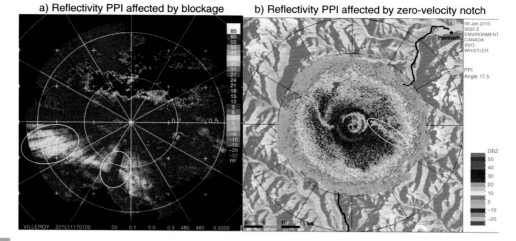

Figure 9.4. Examples of reflectivity data affected by a) blockage and b) an aggressive signal processing algorithm designed to remove ground clutter but which also removed some precipitation with zero Doppler velocity. © Environment Canada, 2012; radar images developed by Paul Joe, Cloud Physics and Severe Weather Research.

9.3.1 Errors in radar measurements aloft

The first step is to carefully measure radar parameters aloft, as close as possible to the surface to minimize eventual problems in extrapolating such values to the surface, but far enough from it to avoid problems due to clutter contamination or beam blockage. As a general rule, the characterization of semipermanent sources of errors of beam blockage has not been incorporated in our algorithms, and images such as the one on the left of Fig. 9.4 are not uncommon, though they rarely make their ways to publications. Using propagation and digital terrain models, one can obtain from simulation where blockage does occur if the source of blockage is a large terrain feature (Kucera *et al.* 2004), but not as well when vegetation or man-made objects are implicated. For the latter, long-term statistics are required, acknowledging the fact that such blockage can change with time. Note that knowing what fraction of the beam is blocked may be insufficient to properly correct for that blockage if there is a significant vertical gradient of reflectivity within the beam (Donaldson 2012).

But the greatest cause of error in radar measurements aloft is attenuation by precipitation of both reflectivity and differential reflectivity measurements. Noticeable at S-band in heavy convection, it can become significant at C-band and quite drastic at shorter wavelengths to the point of losing the echo (recall Fig. 2.12 and supplement e02.4). Accurately correcting for attenuation is challenging because unconstrained correction algorithms become numerically unstable (Hitschfeld and Borden 1954), particularly

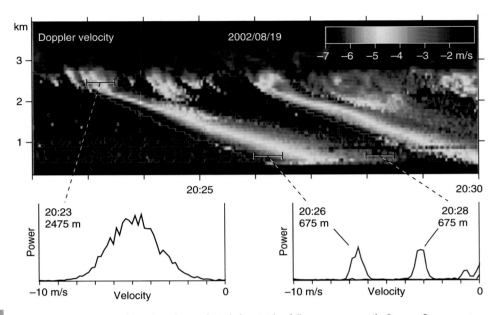

Figure 9.5. Top: Time–height section of Doppler velocity of rain below 2.5 km falling as narrow trails. Bottom: Power spectrum (distribution of echo power with respect to velocity) over 30 s at three locations in a precipitation trail. While drops of all sizes are observed at the top of the trail, different sizes of drops arrive at different times near the surface, leading to a drastic change in drop size distributions (DSDs). Also note the time difference between when and where precipitation is observed aloft, and when it is seen on the ground, differences that may be important when comparing radar and rain gauge accumulations.

when attenuation exceeds 10 dB. If a constraint is available, in the form of differential phase measurements (Testud *et al.* 2000), ground echo strength changes (Delrieu *et al.* 1997), noise intensity changes (Thompson *et al.* 2011), and multiple radar measurements (Chandrasekar and Lim 2008), the algorithm becomes more stable, though none of these constraints have an exact one-to-one relationship with the path-integrated attenuation that we are trying to correct.

To circumvent this problem, in conditions when significant attenuation is expected, some approaches avoid using reflectivity and rely instead on other parameters such as the specific differential phase K_{dp}. K_{dp} is more linearly related to rainfall than reflectivity, though there is still some dependence on drop sizes, and, with some care, has the advantage of being mostly insensitive to what can contaminate reflectivity-based estimates of rainfall such as blockage, or clutter contamination. Unfortunately, it only works well in rain of at least moderate intensity and extent in order to allow clever algorithms to circumvent the large noisiness of K_{dp} estimates. Ultimately, the problem of accurately measuring radar quantities aloft is minimized with the use of a nonattenuating wavelength for long-range applications, or by purposefully choosing a short wavelength and then relying on K_{dp} for short-range applications.

9.3.2 Extrapolating to the surface

While the discussion to follow will center on reflectivity, it applies to all radar-measured quantities. Reflectivity can change considerably with height, especially in cold season precipitation, resulting in considerable biases unless corrective measures are implemented (e.g., Joss and Waldvogel 1990, Fabry *et al.* 1992, Fig. 9.6). In rain formed by the cold rain

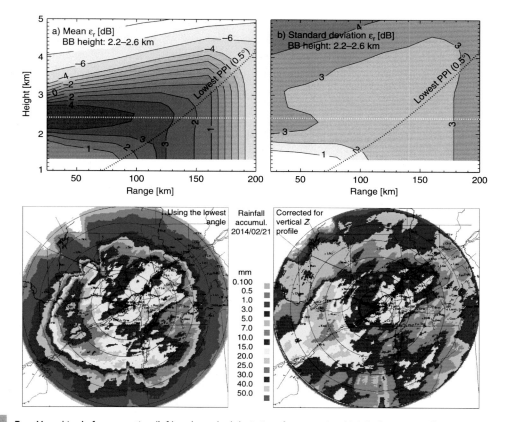

Figure 9.6.   Top: Mean bias before correction (left) and standard deviation after correction (right) of errors in reflectivity at the ground when estimated from reflectivity observed at different ranges and heights for the case where the bright band peak is between 2.2 and 2.6 km based on more than 30 h of profiles, each computed over a 15×15 km^2 area in flat terrain. The dotted line indicates the height and range that a 0.5° PPI would sample. Based on this analysis, without profile correction, overestimation of surface reflectivity is expected from 75- to 170-km range, and underestimation is expected beyond. Republished with permission of the AMS from Berenguer and Zawadzki (2008); permission conveyed through Copyright Clearance Center, Inc. Bottom: Computed 24-h rainfall accumulation up to 240 km range without (left) and with (right) vertical profile correction in an event with a bright band at 2.5 km. Note how the computed rainfall shows a clear range dependence when no correction is applied, and how that dependence diminishes after the correction.

process, in general, estimations aloft are adequate until the radar beam starts to intercept the 0°C isotherm. Then, precipitation is overestimated over a short distance corresponding to the height where the beam goes through the reflectivity enhancement caused by melting, followed by an increasing underestimation beyond as the radar measures gradually weaker echoes from frozen hydrometeors. In snow or in warm rain, radar increasingly under-estimates surface precipitation as range and height increase. The magnitude of such under- and overestimations can be large (Fig. 9.6), particularly in snow or when the bright band is close to the surface.

To a first order, a mapping function is required to transform reflectivity measured aloft to the one that would have been observed at the surface. That mapping function is often computed by assuming a vertical profile of reflectivity to establish what would be observed at different ranges given the antenna elevation and the beam width of the radar (e.g., Joss and Waldvogel 1990). But such a function really works if and only if the distribution of reflectivity aloft has the same width and shape as the distribution of reflectivity at the surface, since the net effect of a correction with a single vertical profile of reflectivity is to add a (range-dependent) constant number of dBs to all echoes independent of their magnitude. In practice, a magnitude-independent single profile of reflectivity rarely exists: there can be very different processes of precipitation formation within the same storm (warm vs. cold rain, convective vs. stratiform) that need to be recognized, for example, using echo intensity and texture (Steiner *et al*. 1995). Even within the same rain formation category, the vertical profile of reflectivity can vary with precipitation intensity (Fig. 9.7). In addition, local terrain can add further complications: in complex topography, more low-level growth is expected than over flatter terrain due to upslope flow. Therefore, care must be taken in properly computing the vertical profiles of reflectivity. The net result is that the further away from the radar and from the surface a radar pixel is, the more difficult it becomes to accurately correct for the change in precipitation between the measurement level and the surface (Berenguer and Zawadzki 2008).

Superposed on the effect of the vertical profile of reflectivity is the effect of the move-ment, or drift, of precipitation between the measurement level and where it will fall on the surface (Mittermaier *et al*. 2004, Lauri *et al*. 2012). The assumption generally being made is that the precipitation observed aloft instantly reaches the surface. Yet rainfall measured 2 km above the surface takes from 3 to 10 min to fall, depending on raindrop sizes; during that time, it also moves horizontally with the average wind speed between those two levels, possibly drifting several kilometers horizontally. If the precipitation process is sufficiently persistent in time that trails can form (e.g., Figs. 4.6 and 9.5), their slope can be used to displace the rainfall pattern horizontally. This may be possible in widespread rain, but not in convective rain where rainfall production evolves too quickly for trails to persist. A proper instantaneous precipitation map at time *t* should therefore ideally use radar data from previous times, corrected for horizontal drift. The error made by not taking this drift into account is small in widespread rain and when averaging precipitation over large basins, but can be significant in convective rain for hourly accumulations and over small areas, such as a very small urban basin or a rain gauge.

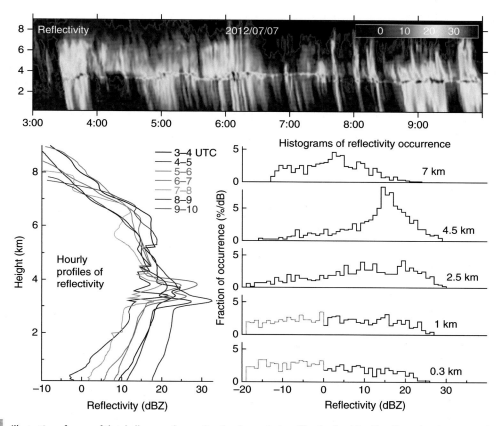

Figure 9.7. Illustration of some of the challenges of correcting for the vertical profile of reflectivity. Top: Time–height section of reflectivity during a storm. Bottom left: Hourly vertical profiles of reflectivity for that event. Bottom right: Histograms of the occurrence of a given reflectivity value at five altitudes (0.3, 1, 2.5, 4.5, and 7 km). Because of insect contamination, reflectivity values below 0 dBZ and at low levels are biased and are illustrated with a dotted line instead of a solid line. The use of a single profile of reflectivity at, say, 4.5 km for the purpose of obtaining low-level values implies the shifting of that distribution down to the surface by an amount corresponding to the mean reflectivity change. The resulting histogram of reflectivity at low levels will consequently be wrong because the vertical profile of reflectivity varies with precipitation intensity, in this case the heavier precipitation experiencing less evaporation near the surface.

9.3.3 Converting to precipitation

After deducing satisfactory estimates of radar-measured quantities at the surface, they must be used to compute precipitation rate, usually with an adequate reflectivity–precipitation, or $Z–R$, relationship. We saw from (3.6) and (3.7) that there is no mathematical relationship between Z and R, reflectivity being overly sensitive to large drops when compared with rainfall. It is therefore essential to determine or estimate the size distribution of precipitation. As a result, long lists of vastly different $Z–R$ relationships have been derived over the years for different areas in different conditions. The observed variability is due in part to the difficulty of comparing data from two data sources, such as radar and gauge measurements, over short time intervals and with different sampling volumes, and partly from the way the relationship between the two data sets is derived (see Appendix A.3 for details on regressions). That being said, Z–R relationships do vary, and that variability must be taken into account. A more rigorous approach for estimating Z–R relationships requires a thorough consideration of the microphysical processes shaping the growth of hydrometeors and the evolution of particle size distributions.

9.3.3.1 Some processes shaping drop size distributions

The rate at which the diameter D of a drop grows is related to how quickly it gains volume or mass, both being proportional to D^3. Let us first consider what it would take to have drops from all diameters grow at the same rate. The rate of diameter change per unit fall distance is constant if

$$\frac{dD}{dz} = \frac{1}{3D^2}\frac{d(D^3)}{dz} = \text{const} \Rightarrow \frac{d(D^3)}{dz} = 3\,\text{const}\,D^2, \tag{9.1}$$

or if the growth rate in mass or volume is proportional to the surface area of the hydrometeor. This is the case for the growth of raindrops by cloud collection, the main process shaping the evolution of light precipitation by the warm rain process, as well as the growth of rimed snow and graupel by liquid cloud collection. If the size distribution of drops is initially exponential $N_i(D) = N_0 exp(-\beta_D D)$, N_0 and β_D being positive constants, then for a relatively small growth ΔD, the final size distribution N_f will be for all but the smallest drops:

$$N_f(D) = N_i(D - \Delta D) = N_0\exp[-\beta_D(D - \Delta D)] = \exp(\beta_D \Delta D)N_i(D), \tag{9.2}$$

implying that the number concentration has increased by a constant factor for all drop sizes. Note that ΔD may vary from one location to the next depending on the amount of cloud accretion that occurs, but is the same for a given drop population. If that is the case, then both Z and R increase by the same fraction, leading to a linear relationship between Z and R, a phenomenon actually observed in drizzle (Fig. 9.8).

This exercise demonstrates how a process, or a combination of processes, leading to a size-independent fractional change of $N(D)$ results in a proportionality between Z and R. As just illustrated, it is observed in light warm rain but it also occurs with the highest rainfalls. When drops collide, coalescence or breakup can occur, depending on drop sizes and the energy of the collision. If enough collisions between two sizes of drops D_1 and D_2 occur, the distribution of drops resulting from those collisions is predictable. By summing over all D_1 and D_2, it is possible to determine how each drop size is transformed by collisions. After a large number of collisions and breakup, all memory of the initial DSD is lost, and the DSD converges such that the number of drops of each size is linked by a constant to the number of drops of any other size. This is known as the equilibrium DSD caused by coalescence and breakup (Valdez and Young 1985; Zawadzki and De Agostinho Antonio 1988). It is observed near the center of tropical

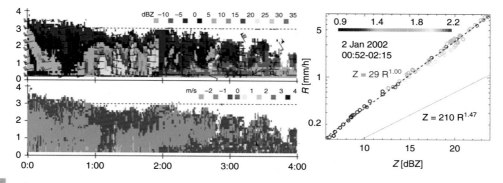

Figure 9.8. Left: Time–height section of reflectivity (above) and of Doppler velocity (below) in a warm rain event in Montreal, Canada. Right: Reflectivity–rainfall relationship derived by a disdrometer at the surface. The symbol color corresponds to time (in hours since midnight). In this example, warm rain drops are very small and fall slowly. The resulting $Z–R$ relationship, $Z = 29R$, is linear and very different from the climatological relationship observed in Montreal ($Z = 210R^{1.47}$). Images courtesy of Isztar Zawadzki.

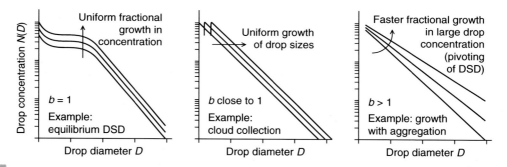

Figure 9.9. Illustration of how DSDs and the exponent b of a $Z–R$ relationship will vary for different dominant growth processes. Larger b is obtained when the concentration of large drops increases faster than that of small drops, leading to an apparent pivoting of the DSD. If it pivots around $D = 0$ such as for the Marshall–Palmer DSD, we obtain $b \approx 1.5$.

thunderstorms where high rain rates and numerous drop collisions occur. When such an equilibrium is reached, any additional process leading to growth or decay of precipitation redistributes its contribution to all drop sizes, thus maintaining the proportionality between Z and R.

After these two examples of proportionality between Z and R, one may wonder how "standard" $Z–R$ relationships such as the $Z = 300 R^{1.5}$ shown in (3.9) occur. To obtain a relationship $Z = a R^b$ with the exponent b greater than 1, the number of larger drops must increase fractionally faster than the number of smaller drops as the rainfall rate increases (Fig. 9.9). This may happen in growth by accretion as seen above, when the diameter change ΔD becomes too large to ignore the deficit in the number of small drops. However,

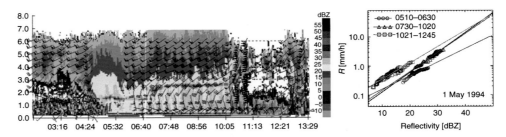

Figure 9.10. Left: Time–height section of reflectivity in changing rain. Early in the precipitation event (05:10–06:30), a thick region of high-reflectivity snow is observed where significant aggregation can occur; later (07:30–10:20), a more usual cold rain profile is observed; finally (10:21–12:45), warm rain occurs. Right: Average reflectivity–rainfall relationship observed for the three periods. Drizzle, with its small drops, has low reflectivity given the observed rainfall, while the rain originating from the thick layer of aggregating snow has high reflectivity given the observed rainfall as well as a higher b in the $Z = aR^b$ relation illustrated by a shallower slope on the reflectivity–rainfall plot. The image on the right is republished with permission of the AMS from Lee and Zawadzki (2005a); permission conveyed through Copyright Clearance Center, Inc.

it most commonly occurs as a result of the combined effect of aggregation or coalescence with other growth processes.

Aggregation and coalescence diminish the number of small particles and increase the number of large ones, thus also increasing reflectivity; breakup does the opposite. When taken in isolation, none of these processes significantly affect precipitation rate. Hence, since they change Z but not R, aggregation, coalescence, and breakup primarily shape the prefactor a of $Z = a R^b$ relationships. Snow is more effective at aggregating than rain is at coalescing; this partly explains why the vertical gradient of reflectivity with height is high in snow and low in rain, as well as why cold rain of a given intensity has larger drops than warm rain. When aggregation can occur over a deep nearly isothermal layer at temperatures near 0°C, huge snowflakes just above and very large raindrops just below the melting layer are observed, resulting in a large reflectivity Z for a given rainfall rate R (Fig. 9.10).

So what shapes b? It is the covariance, or the tug-of-war, between particle size growth processes like cloud collection, and particle size redistribution processes such as aggregation. If, for example, for every incremental change in growth, there is a large change in aggregation, b will be large. Otherwise, b will be smaller, though still greater than 1. Large changes in aggregation resulting from small changes in crystal growth may occur when diffusional growth happens at temperatures favoring the growth of dendritic or needle crystals, as these crystals are most effective at attaching themselves to other crystals in the event of a collision, while others such as plates are not (Fig. 9.10).

Other processes shaping DSDs are discussed in Rosenfeld and Ulbrich (2003). But the conclusion of this discussion is that a more intelligent choice of a Z–R relationship can be made when the dominant microphysical processes shaping the growth of precipitation particles are known. These processes shape the climatological Z–R relationship observed

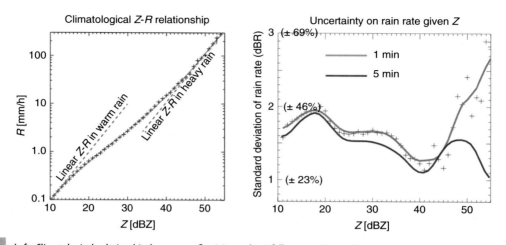

Figure 9.11. Left: Climatological relationship between reflectivity and rainfall rate in Montreal, Canada, obtained from disdrometer data. At both very light and very heavy rain rates, Z tends to be linearly related to R while a power–law relationship would better describe the intermediate rates. Right: Uncertainty in dB (rain rate) on 1-min (green line) and 5-min (orange line) averaged DSDs as a function of reflectivity. On the left axis, the unusual dB (rain rate) units are converted in fractional uncertainty in parenthesis. For 5-min averaged DSDs, smaller fractional uncertainty on Z–R conversions occurs at higher rain rates. Images courtesy of Isztar Zawadzki.

in any given area (Fig. 9.11): in continental and temperate climates, where the DSD is primarily shaped by cold rain processes, drops tend to be larger for a given rainfall rate than in maritime and tropical regions, where warm rain processes become more important. But variations around that climatology do exist and are primarily a function of the dominant precipitation growth processes for the storm event considered.

Although this description is somewhat complex, much of the information required to determine dominant growth processes could be retrieved by studying how the echo strength and polarization characteristics vary with height. However, a proper quantitative analysis of how to determine the optimum prefactor a and exponent b of a $Z = a R^b$ relationship has, to my knowledge, not been done, which is why a more definite guidance on how to choose the best Z–R relationship for a particular event cannot be given. And such a work may become obsolete, or conversely easier to achieve, with the growing preponderance of dual-polarization radars that provide estimates of mean drop size.

9.3.3.2 Polarization-based approaches

Polarization-based measurements such as the differential reflectivity and propagation phase delay can provide additional information on target properties that can be used to estimate precipitation. As we saw in Chapter 6, we can use three additional clues to estimate rainfall: (1) the copolar correlation coefficient ρ_{co}, together with other radar quantities, helps us determine if we are really dealing with rain alone or if the other radar quantities may be contaminated by echoes from other targets, (2) the differential reflectivity Z_{dr} can be used to estimate a mean drop shape from which a mean drop size can be derived. Given a constraint on the mean drop size of the distribution, the range of allowable Z–R relationships is considerably reduced, provided Z_{dr} can be estimated accurately; and (3) the differential propagation phase delay K_{dp} can also be used to provide another estimate of rainfall that has the great advantage of being unaffected by calibration and attenuation problems. However, claims that K_{dp} is independent of drop size are exaggerated: while it is less sensitive than Z to DSDs, there is no one-to-one relationship between K_{dp} and rainfall, and a Z_{dr}-based correction is needed except at the highest rainfall when Z, R, and K_{dp} all become linearly related thanks to the equilibrium DSD. The main challenge in using K_{dp} remains its noisiness, being a quantity that is difficult to estimate accurately, especially for weak rainfall. K_{dp}-based estimates are thus best suited for longer duration accumulation and larger basins.

Nevertheless, experiments in estimating rainfall using dual-polarization approaches reveal that the conversion to rainfall can be done more accurately than with using reflectivity alone (Ryzhkov *et al.* 2005; Fig. 6.17). On the technical side, the key challenge is to obtain a Z_{dr}, or an independent drop size estimate, that is free of bias. On the science side, questions remain as to whether the average axis ratio of raindrops (Fig. 6.1) can be affected by turbulence or drop collisions, and hence become itself a function of DSDs, which would introduce another source of uncertainty to all polarization-derived relationships. For these reasons, and because Z_{dr} can be affected by attenuation, polarization-based rainfall estimates used operationally tend to limit their reliance on Z_{dr} despite the great value it could add if it could be measured accurately. Instead, a blend of Z, Φ_{dp}, and K_{dp} is being favored where Φ_{dp} differences along long paths are used to correct Z for attenuation and calibration errors, while a combination of Z and K_{dp} provides local rainfall estimates depending on rainfall intensity. But research continues, and with growing experience using dual-polarization measurements, better or more robust methods to obtain rainfall estimates will be derived.

9.3.3.3 Snowfall estimates

The major remaining challenge is snowfall estimation. In rain, a mean drop size, and hence a mean fall speed, can be deduced from measurements of differential reflectivity, thus helping to fine-tune the conversion of reflectivity into rainfall. In snow, there is no one-to-one relationship between shape/orientation and snowflake size, although there is a tendency for larger snow aggregates to be more spherical than individual crystals or smaller aggregates. Moreover, snow fall speed is mainly controlled by the density of the snowflake, the latter in turn being more a function of the degree of riming rather than of snowflake mass. Therefore, polarization data cannot be easily used, and we are forced to primarily rely on reflectivity and its vertical profile to infer snowfall intensity. Equivalent reflectivity to snowfall, or Z_e–S, relationships do exist, but they tend to be more uncertain than Z–R relationships because of the many possible processes shaping the Z_e–S relationships as well as the difficulty in accurately measuring snowfall rate with

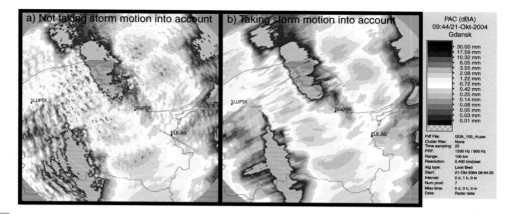

Figure 9.12. Comparison of two 1-h accumulations, one made without taking storm motion into account (left) and one by taking storm motion into account (right). When motion is not taken into account, artificial rainfall structure is created at locations where individual cells were observed at different times. Images used with permission of Selex ES GmbH.

in situ sensors to validate such relationships. They also vary with temperature: at cold temperatures, snow falls as smaller individual crystals; as temperature warms and crystals grow, they start to aggregate. Near 0°C, snow-equivalent reflectivity above the bright band is, on average, 1–2 dB weaker than rain below the bright band for a wide range of rainfall reflectivity (Fabry and Zawadzki 1995). On the basis of this observation, a Z_e–S relationship of $Z_e = 200S^{1.5}$ should on average provide adequate snowfall estimates near 0°C. But given the strong reflectivity gradient in snow above the bright band caused primarily by aggregation, it would be expected that the prefactor a would decrease steadily from 200 near 0°C to something closer to 100 at −10°C. What is clear is that we have yet to find the best method to use radar data to estimate snowfall. For more information on snowfall estimation by radar and its challenges, consult Saltikoff *et al.* (2014).

9.3.3.4 The final accumulation

Deriving precipitation rate at the surface is not the last step. It must then be properly accumulated in time. The easiest approximation is to assume that the rain rate does not change between radar observations but such an approximation leads to accumulation maps that have unrealistic features (Fig. 9.12). Not unlike the effect of precipitation drift mentioned previously, the use of such a simplistic approximation contributes significantly to errors over very small basins and to the scatter in radar-gauge comparisons (Fabry *et al.* 1994), but has smaller effects on longer duration accumulations over large basins. Most radar processing systems, though regrettably not all of them, now take into account the motion of storms and the change in their intensity between successive maps when computing accumulations.

9.3.4 Basin smoothing; gauge adjustments

One mitigating factor and one adjustment technique that contribute to reduce errors in radar-based precipitation estimates are the smoothing behavior of basins and the adjustment of radar accumulation with gauges. If the main goal of obtaining precipitation maps is for river management, the most important correction to be made is the removal of any basin-wide bias in the radar estimates. Once this is achieved, radar will capture well the spatial variability of rainfall within the basin as well as the average rainfall over the entire basin resulting in useful forecasts of river level.

Within this context, errors can be classified into three categories:

a. Bias errors and errors with correlation distances much larger than the basin size, implying that the sign and to some extent the magnitude of the error do not significantly vary inside the basin. It is essential to correct these errors, but because their magnitude is nearly uniform, they are also the easiest to fix via a gauge adjustment. These include errors due to calibration and reflectivity–rainfall conversions.
b. Errors with correlation distances much smaller than the basin size, implying that the sign and magnitude of the error vary rapidly within the basin. Errors such as those due to precipitation drift or a poor accumulation algorithm often fit in that category for larger basins. While the correction of these errors may look complex, they do not significantly affect the net river-level prediction because they largely cancel each other. To a large extent, they can be ignored.
c. Errors with correlation distances comparable to the size of the basin: Errors such as those caused by attenuation and the vertical profile of reflectivity belong to this category. They generally have complex geometries and vary considerably from one section of the basin to the other. While the average error could be corrected by an external data source such as rain gauge measurements, there will remain a considerable amount of error left after adjustment that will affect river-level forecasts.

Note that the size of the basin dictates which error fits in which category: an urban or mountain creek basin is more sensitive to smaller scale errors than a larger rural basin generally found at lower elevations. Because errors belonging in the third category cannot be easily corrected by gauge adjustments or by the smoothing effect of basins, it is hence essential to correct for them at an earlier stage by properly taking into account the nature and geometry of the error.

Consider, for example, the errors shown in Fig. 9.4a. If the gauge used for the correction is located outside the blockage, the reduced rainfall due to this blockage will not be corrected; if the gauge is under the blocked area, the entire rain field will be increased, leading to major errors outside the blocked area; and any attempt to use a combination of gauge values without taking into account the geometry of the error will result in a rainfall field with errors having long correlation distances that will introduce major errors in water-level predictions.

These details need to be considered when developing strategies for adjusting radar-derived precipitation maps with rain gauges. The simplest and most robust method is to compute an average gauge-to-radar, or *G–R*, ratio, and use it to correct the mean bias for the entire field. This has proved to be the most accurate and safest way to adjust radar maps, and it works best when errors having unusual geometries have been properly corrected. The temptation is then great to use multiple gauges to build a space-varying correction field, but in practice, the results of such efforts have generally been unsuccessful for reasons discussed in the previous paragraph: since many radar errors follow a spherical geometry (either having the shape of spokes aimed at the radar or circles with the radar at its center), algorithms that assume that error patterns are isotropic will cause distortions in the error correction. Therefore, any gauge adjustment should always be the very last step in the radar accumulation process, and one that should be done very carefully.

9.4 Uses

Despite their shortcomings, precipitation accumulations derived from radar have a variety of uses that change from country to country. Shorter duration (1–3 h) and longer duration (6–24 h) accumulations of precipitation are routinely generated by radar data processing systems to help forecasters monitor the potential for floods. These are also used as input to hydrological models, systems that combine precipitation and terrain information to forecast the flow of rivers or of sewers. In postanalysis, radar and gauge accumulations are often combined to obtain analyses of precipitation that have a variety of uses from the validation of precipitation forecasts from weather forecasting models to precision agriculture.

Compared with gauges and many other instruments, radar has an additional advantage for flood forecasting, especially for smaller basins: not only is it capable of quantifying past precipitation amounts, but it also has the potential to estimate precipitation accumulations in the near future.

10 Nowcasting

10.1 Nowcasting needs and approaches

Almost everyone who has looked at a real-time animation of radar imagery has wondered: What is the weather going to do next and where is it heading? Will it reach my location? If yes, when, and for how long? Will the weather be severe? These are the questions that nowcasting systems and techniques are designed to answer.

For a variety of applications, very short-term forecasts of a few minutes to a few hours are needed. These may include the determination of the future track of a particularly severe storm for warning purposes, or the estimation of the amount of additional precipitation that will fall in a given area in the next few hours. Since weather varies rapidly in unexpected ways, such forecasts must be recreated often, typically several times per hour. Because numerical weather forecasting models must first wait several minutes for all the needed data to be available and then produce a correct analysis at the initial time before finally generating their forecast, they cannot at present satisfy our needs for very frequent forecast updates. Furthermore, they often do not perform very well for short lead times (Fig. 10.1). Therefore, there exists a niche for simpler and faster forecasting approaches.

Etymologically, nowcasting comes from the contraction of the words "now" and "forecasting." It refers to techniques dedicated to make forecasts over relatively short periods, generally with a lead time within 12 h. These are generally less complicated and designed to function more efficiently at short time scales than the traditional weather forecasting based on numerical modeling covering large portions of continents. For example, when we make a short-term forecast based on the pressure tendency of our home barometer, we are in essence performing a nowcast. Nowcasting techniques are particularly suited to use data from remote sensors such as radar and satellite. Radar data can be used in different types of nowcasting techniques that include the following:

a. Precipitation nowcasting, where the time history of precipitation fields is being used to estimate where precipitation will occur in the future.
b. Severe weather nowcasting, where areas of dangerous weather are first recognized and then their expected trajectory estimated.
c. Mesoscale forecasting systems, where the radar data are being assimilated with other data into a numerical weather forecasting system that then generates a short-term forecast.

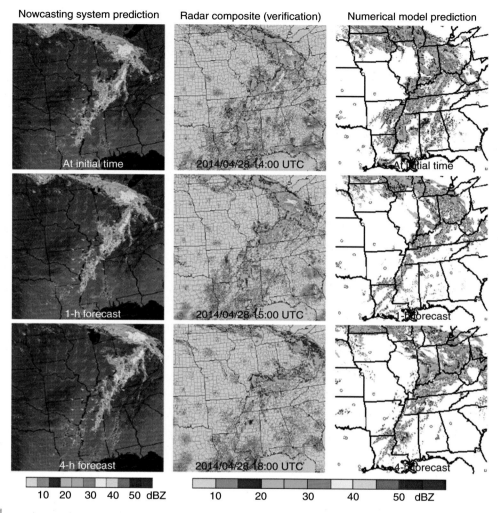

Figure 10.1. Comparisons of reflectivity patterns over a 1920 × 1920 km area made by a nowcasting system (MAPLE, Turner *et al.* 2004, left column), radar observations (middle column), and predictions made by a numerical prediction model (HRRR, Alexander *et al.* 2010, right column) at the initial time (top row), and for 1-h (middle row) and 4-h (bottom row) forecasts. Note how (i) the model precipitation at the initial time is slightly different from the radar observations, (ii) the model rapidly creates and dissipates precipitation in some areas (e.g., near the center and to the south of the domain for the 1-h forecast) where the dynamical support from the model is incompatible with the initial precipitation field provided from observations, and (iii) how the nowcast-based prediction cannot properly capture precipitation evolution for longer lead times. As a result, nowcasts perform better for short lead times as they better capture initial conditions, while numerical forecasts do so for longer lead times as they better handle storm evolution. Images on the left column are courtesy of Alamelu Kilambi; those on the center column are © 2014 UCAR, used with permission; those on the right are courtesy of NOAA.

The first two sets of nowcasting approaches are often integrated into radar data processing systems used in weather or hydrological forecasting offices. The last approach requires considerably more computing power and is typically run more centrally. In this chapter, the bases of all three approaches are introduced.

10.2 The basis for simple nowcasting systems: extrapolation

What is required to obtain a very short lead forecast? Consider first an extreme example: the 30-s forecast. For such a short forecast lead time, an extremely simple yet very effective approach is to assume persistence, that is, that the weather being presently observed is the one that will occur 30 s later. What is hence needed for such a short-term forecast is an excellent observation, and the quantities that can be forecasted are thus limited to what can be observed as well as to what can be directly inferred from these observations. Radar provides unique observations of precipitating systems and of the radial wind patterns within them at frequent time intervals (a few minutes) and at fine spatial resolution (~1 km); therefore, nowcasting systems using radar data alone can only forecast the precipitation patterns and severe weather detected by radar. Note that in practice, a 30-s forecast is not very useful and is more difficult to achieve with radar than it first appears, but it serves as a useful example to illustrate nowcasting concepts.

If the forecast time is now extended to 5 min, persistence starts to fail, especially when trying to forecast precipitation intensity and convective weather, primarily because of storm movement. For 5-min forecasts, a nowcasting system must take into account that movement while storm evolution can still be ignored under most circumstances. The forecasting approach is therefore first based on the determination of past storm movement followed by its extrapolation in the future. This is basically what we do when we observe an animation of radar images to determine where the storm will move to. Systems based on this approach are sometimes said to assume Lagrangian persistence, meaning persistence along the direction and speed of movement of storms. Essentially all nowcasting systems use extrapolation based on storm movement to determine the future position of existing storms or threats.

For forecasts beyond 5 min, systems relying on Lagrangian persistence alone make perceptible errors because storms evolve and change direction. A key fact to remember is that smaller scale weather features evolve faster than those that are larger in scale. Revisiting Fig. 8.1, it can be expected that isolated thunderstorms 10 km in size will undergo considerable evolution over 30 min, while storm clusters 100 km wide will do the same in a couple of hours. Because this complicated evolution is difficult to forecast, especially for simple systems based on simple rules, it sets a limit on the maximum lead time for which nowcasting systems can forecast weather phenomena of different sizes. In other words, as lead time increases, the probability that a currently observed threat will still exist at that lead time diminishes rapidly. Despite what may seem to be the case by casually looking at radar animations, the storm cell that is observed 60 km away and expected to arrive in an hour will generally decay before reaching your location; what is more likely is that another storm not yet formed but from the same convective complex will be the one bringing precipitation and severe weather, and probably not in exactly 1 h.

Finally, some parts of the world are more fortunate than others when it comes to nowcasting. In relatively flat terrain, and away from warm seas and oceans, large-scale patterns of precipitation often originate from regions upstream. Hence, even if individual cells cannot be forecasted for a long time, large precipitation areas can still be effectively forecasted by extrapolation. In other parts of the world, such as in tropical coastal regions or near mountainous areas, much of the precipitation is locally generated. It then becomes essential for nowcasting systems designed for these areas to undertake the complex task of predicting storm initiation. This reality has led to the development of nowcasting systems with two very different philosophies, namely one where the forecast is dominated by variations on the extrapolation of existing systems, and another where considerable efforts are spent to determine where and when storm initiation may occur.

10.3 Precipitation nowcasting

Because different types of users have different forecast needs, there exists a variety of nowcasting systems. Systems designed for the nowcasting of precipitation tend to be field oriented: the precipitation field is considered as a single pattern that can move and be deformed with time. The first step in nowcasting is to determine the extent of the deformation of that entity by computing a set of velocity vectors, one for each tile of a fixed size (Fig. 10.2), and use them to predict future echo movement, as was done in Fig. 10.1. Different methods exist to compute velocities rooted in the maximization of cross-correlation of patterns (Bellon and Austin 1978, see also Appendix A.4 for background information on correlation) or based on the optical flow constraint but applied to smoothed echo fields (Bowler et al. 2004). When multiple vectors are needed, the procedure is modified such that all vectors are computed simultaneously subject to a consistency constraint that stabilizes the result of the motion calculation as in Germann and Zawadzki (2002).

In addition to being deformed, the forecasted precipitation pattern can have its intensity modulated to take into account growth and decay. This has however proved to be a frustrating endeavor as trends in precipitation intensity are often short-lived and constantly change (Tsonis and Austin 1981, Radhakrishna et al. 2012). Only the very large-scale trends such as those forced by the diurnal cycle may have some predictability, but these trends vary considerably from location to location and with the seasons. As a result, most precipitation nowcasting systems do not include growth and decay in their prediction, and those that do perform better with decay than with growth.

In reality though, precipitation does evolve with time, and thus the predictions of nowcasting systems necessarily incorporate an increasing error with lead time that should be evaluated. Hence, increasingly, precipitation nowcasting systems attempt to compute

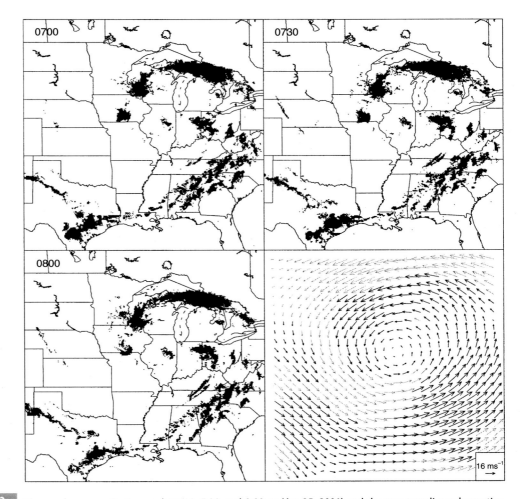

Figure 10.2.   Three radar composite images (at 7:00, 7:30, and 8:00 on May 25, 2001) and the corresponding echo motion field obtained by tracking reflectivity patterns. On the images, only the areas with reflectivity greater than 10 dBZ are shaded. Computed vectors are most reliable in regions with precipitation echoes (black arrows), whereas far away from any precipitation they must be interpreted with care (light arrows). Republished with permission of the AMS from Germann and Zawadzki (2002); permission conveyed through Copyright Clearance Center, Inc.

expected uncertainties on their forecasts. To make this assessment, a set of possible predictions are generated, known as an ensemble of forecasts; the goal is then to use the width of the distribution of precipitation forecasts at each point to quantify forecast uncertainty. In practice, the ensemble of forecast is made as follows. The original nowcast prediction is used as the starting point. At each forecast time, the smaller scale patterns that are expected to have evolved considerably are smoothed to keep only the larger scale patterns that are predictable. This implies that, as lead time becomes larger, the rainfall patterns that are judged to be predictable become increasingly smooth. Using this smooth rainfall pattern as a basis, for each ensemble member, random patterns are added that have the spatial and statistical properties of the echoes that have been removed by smoothing. A different random pattern is added to the base map for each ensemble member. The result is an ensemble of forecasts that will have in common the predictable patterns, but will differ in the patterns that are expected to be unpredictable.

The spread in each forecast provides information on forecast uncertainty and also on the probability of occurrence of different outcomes, a key piece of information required to issue warnings or to make decisions on how to manage a river basin. More details on the implementation of such a system can be found in Seed (2003) and Berenguer *et al.* (2011).

The forecast skill of precipitation nowcasting systems depends on the quality and spatial extent of input radar data. When applied on data from single radars, useful nowcasts can be made up to 2 h, the lead time being limited by the maximum useful range of the radar data. When applied on radar composites, the range of useful prediction can extend to approximately 6 h. Poorer radar quality hampers nowcast accuracy at all lead times; even with good quality radar data, an important fact to remember is that for 1-h predictions, approximately half of the errors in rainfall forecasts are due to errors in the conversion from radar-measured fields to rainfall patterns at the initial time (Fabry and Seed 2009). Comparisons made between forecasts by nowcasting systems and numerical models show that, at present and on average, nowcasts perform better than mesoscale model forecasts up to 3 h (Fig. 10.3). Precipitation nowcasting systems are in use in many countries in either meteorological or hydrological services but tend to be more common in countries where flooding is the main weather hazard. Pierce *et al.* (2012) provide more details on existing systems and their principle of operation.

10.4 Severe weather nowcasting

While precipitation nowcasting systems were designed to predict future precipitation fields and their uncertainty, systems dedicated to the nowcasting of severe weather predict the outcome of threats. Their role is to help forecasters decide whether to broadcast a warning or not (recall Section 7.3.3). These systems use a variety of algorithms to detect regions of severe weather from real-time radar data, in particular threatening thunderstorms, hail cores, mesocyclones, etc. The set of threats being forecasted varies depending on whether the nowcasting system is designed for aviation users, weather offices, or civil protection bureaus. The basis of these systems is the same as for precipitation nowcasts: many of the threats observed at the current time have likely been detected for the same storms in the previous volume scans; if they can be tracked, their position can be forecasted in the near future.

Because each of these threats is generally associated with a very specific area with boundaries defined by objective rules, they are often portrayed as objects. Each type of

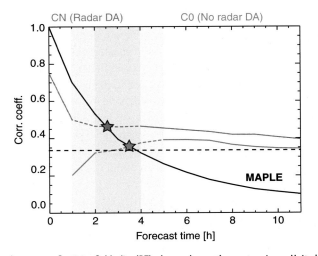

Figure 10.3. Correlation coefficient between reflectivity fields (in dBZ) observed over the conterminous United States and those predicted by a nowcasting system (MAPLE, black line), and a mesoscale model (Storm Scale Ensemble Forecasting System, Xue *et al.* 2008) initialized with (orange) and without (cyan) radar data assimilation. The star indicates the crossover point beyond which predictions by numerical modeling become better than that from the nowcasting system. Shaded areas show when the difference between model and nowcast skill varies from one event to the next, orange-shaded areas contrasting nowcasting and model forecasts with radar data assimilation, and cyan-shaded areas contrasting nowcasting and model forecasts without radar data assimilation. Image courtesy of Madalina Surcel.

objects has properties that can be characterized: location and magnitude of the peak echo intensity or rotational velocity, area or volume of echoes above a threshold, etc. Each of these objects can also be tracked independently from one another and from the precipitation field itself. The nowcasting of severe weather therefore consists in (a) building a list of threat objects at the present time and of their characteristics, (b) if possible, associate each object of this list with the corresponding object of the previous scan, and (c) use the past movement of each object to project a possible trajectory in the future (Fig. 10.4). Examples of descriptions of the approaches used to identify and track thunderstorm cells can be found in Dixon and Weiner (1993) and Johnson *et al.* (1998). The task of associating threats at the present time with those from the past is challenging because threatening weather evolves rapidly in time: for example, a severe storm may split into two, or a mesocyclone may vary with an intensity that oscillates around the detection threshold. The tracking of these objects is therefore more difficult if the time between volume scans is longer.

Despite the challenges, severe weather nowcasting systems are in common use in many weather services, each system having a slightly different emphasis as severe weather forecasting needs change from country to country. A partial list and brief description of systems in use circa 2010 are presented in Joe *et al.* (2012). Because of their focus on severe weather, forecasts are usually generated for only up to 1 h lead time, considered as the longest lifetime of most convectively induced circulations. Most of these systems predict

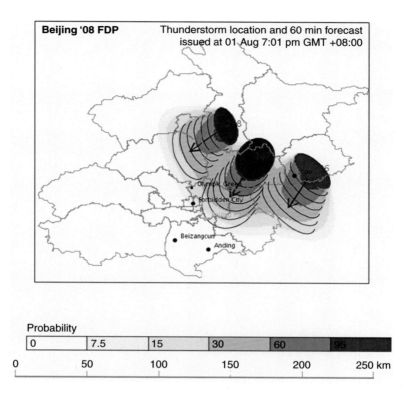

Figure 10.4. Example of a nowcast of thunderstorms made by a severe storm nowcasting system, the Thunderstorm Interactive Forecast System (Bally 2004), for the 2008 Beijing Olympics demonstration project. The location of each severe thunderstorm is identified with a number, and the shape of the cell is synthesized by a full ellipse. The arrows and partial ellipses illustrate the average and the range of possible tracks that these thunderstorms can take while color shading indicates the strike probability. From Joe *et al.* (2012), © Copyright 2012 InTech, licensed under Creative Commons 3.0 rules.

the motion of existing threats and do not attempt to forecast the appearance of new threats such as the initiation of convection.

A notable exception to that description is the Auto-Nowcaster (Mueller *et al.* 2003; Roberts *et al.* 2012). In the foothills of the Rocky Mountains (Colorado), the region for which it was originally designed, essentially all the convective weather is initiated locally (Fig. 7.3), rendering unsuitable approaches based on the tracking of existing phenomena. The successful prediction of convective initiation relies on a combination of the detection of radar-observed signatures such as reflectivity fine lines (Fig. 5.14) and their extrapolation, satellite observations of growing cumulus, and atmospheric stability information from models (Fig. 10.5). Forecaster input is allowed to help the system detect the presence of reflectivity fine lines that are otherwise difficult to detect automatically and to help the system determine which signatures are more meteorologically important. Thanks to all these inputs, it has shown some skill in forecasting initiation, especially when convective forcing is well organized.

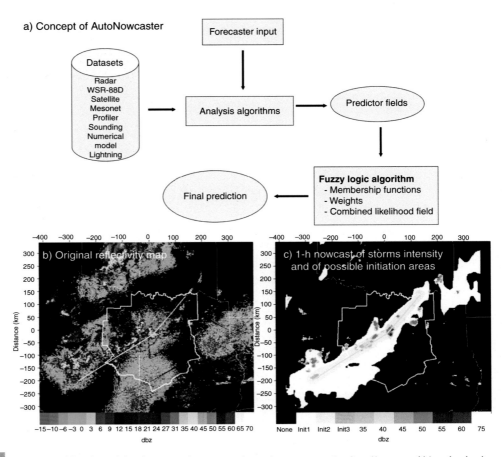

Figure 10.5. a) Conceptual flowchart of the elements and process used to make a nowcast using Auto-Nowcaster. b) Low-level radar reflectivity mosaic at the time the forecast is made. A yellow curved line illustrates where the convergence boundary is expected to be in 1 h. c) Final 1-h convective storm nowcast field. Gray-shaded regions represent three levels of increasing likelihood of storm initiation, while color shades show the expected location and intensity for the existing storms in (b). The conceptual flowchart is courtesy of Rita Roberts; the two bottom images are republished with permission of the AMS from Roberts et al. (2012); permission conveyed through Copyright Clearance Center, Inc.

10.5 Mesoscale modeling after radar data assimilation

In order to extend the forecast horizon of traditional nowcasting systems and also as an attempt to exceed their skill, considerable research effort is being directed to the development of forecasting systems that assimilate radar data. This approach is based on the idea that numerical weather prediction is the most accurate method to forecast weather, because it is the only one that incorporates all the key atmospheric processes. But weather prediction is what physicists refer to as an initial value problem, for which it is necessary to have the most accurate knowledge of the current atmospheric state, small initial errors growing rapidly with time. To minimize these errors, a method is needed to optimally combine all the available information about the present weather. The approach, and by extension, the set of techniques used to determine the present state of the atmosphere, is known as data assimilation. These techniques were pioneered for global-scale models, but mesoscale and convective-scale models now commonly use them. At these smaller scales, radar provides unique information about wind and precipitation patterns; radar data assimilation therefore is essential for short-term forecasting purposes.

Sun and Wilson (2003) introduce the most significant radar assimilation techniques, while Lewis *et al.* (2006) provide the underlying theory and the notation used. They are all based on the combination of two sets of information: the existing knowledge of the state of the atmosphere, known as the "background" or as the "a priori," and the new knowledge gathered from observations using radar and other instruments. The goal of data assimilation is to optimally combine these two sets of information to determine the best possible estimate of current conditions, known as the "analysis" or as the "current state." In order to achieve that optimal combination, data assimilation relies on additional sources of information such as the expected errors on the background and on observations, on algorithms that simulate observations from a given model field, and sometimes on other constraints such as field smoothness or model equations to be respected. The main difference between different data assimilation methods rests on how this optimum combination is achieved and the extent with which nonlinearities must be taken into account. Figure 10.6 illustrates both how data assimilation fits in the context of weather forecasting and how it is generally implemented.

To minimize the errors on the analysis and thus ensure a successful data assimilation, it is essential to have the following:

a. Good a priori information from the background: The background, or the a priori, refers to the original estimate of the present state of the atmosphere before considering new information. It generally comes from a previous forecast. Because radar will provide new observational constraints primarily on precipitation and on one component of the wind, the information about other fields such as temperature, humidity, and tangential wind components must come from that background. These unobserved fields are however what ultimately cause storms to occur. Hence, if the background does not have the correct dynamics or thermodynamics to sustain the radar observed fields, whatever modifications will be made to the precipitation fields by adding radar observations will tend to be quickly forgotten as shown on Fig. 10.1. A successful data assimilation thus requires a background as close to reality as possible. This may not be achievable when data assimilation cycles are first started, but it is hoped that the background will improve as time progresses and the data assimilation process is repeated with new observations.

b. An accurate background error structure: The background is expected to have errors, but an excellent knowledge of that uncertainty or of the expected errors on that information will permit an optimal combination of the available information. But because errors in one quantity at one location, say vertical air velocity 1 km above the surface, are often

a) Data assimilation within weather forecasting

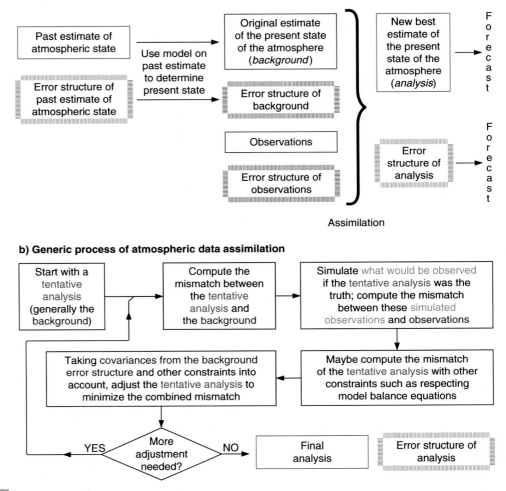

b) Generic process of atmospheric data assimilation

Figure 10.6. Flowcharts illustrating a) the context within which assimilation is performed and b) the process of atmospheric data assimilation. In this figure, the process of data assimilation was separated into four steps to more clearly illustrate what is being achieved; in practice, data assimilation algorithms combine these four steps in one single calculation.

caused by errors in other quantities at other locations, such as wind convergence or temperature below that height, we need to describe not only how large the expected errors are for each quantity at each location, but also how each of these errors is related to every other error in the background fields. The task of describing the complicated structure of the errors of the background rests on the background (or a priori) error covariance matrix. It codifies how large each error is and how correlated every error is expected to be with errors in other quantities and locations. To a large extent, the success of data assimilation rests on that error description: it is the main pathway by which the data assimilation system recognizes that if a background field such as precipitation does not correspond to an observed pattern, then the unobserved updraft, temperature, and humidity fields in the background must also be wrong, and it uses the background error covariance matrix to estimate how to correct all fields in a way that respects those expected error relationships. If these error relationships are missing or poorly estimated, it becomes very difficult for the data assimilation to constrain the unobserved fields. Additional constraints based on common sense or model physics, such as "saturate with humidity any updraft where echoes are observed," can be used as crutches to help error information travel from one field to another, but they cannot be totally effective. Depending on the data assimilation approach used, the background error covariance matrix can be preset in advance or dynamically calculated from an ensemble of estimates of the present atmospheric state (Bannister 2008). Its improper determination remains the main reason why many data assimilation efforts do not perform as well as anticipated.

c. Unbiased and uncontaminated observations: The theory and approaches used in data assimilation rely on the assumption that the information used may have errors, but that the mean value of these errors at every location is zero, or that there is no systematic bias anywhere. Therefore, imperfect data quality, such as a calibration bias on reflectivity or a Doppler velocity bias introduced by ground clutter and ground clutter removal algorithms, is detrimental to data assimilation efforts. Contrary to human users who can learn to tolerate bad data but are very distressed when there is no information over a given area, data assimilation algorithms can deal with a lack of information but create very unrealistic analyses if the data are contaminated. As a result, when data quality at one location is doubtful, it is preferable to remove that new information rather than possibly providing a biased value. Note that "no information" is very different from "no echo," as the absence of an echo is useful information, one that is very valuable for eliminating from the analysis nonexisting precipitation.

d. An accurate algorithm for simulating radar observations: To be able to use real observations as constraints, the data assimilation system must be able to simulate what would have been observed by radar or other instruments for any given atmospheric state in order to verify how compatible true observations are with that state. This observation simulation is done by an algorithm called an "observation operator," also sometimes referred to as a "forward." Examples of observation operators for radars can be found in the references in Thompson *et al.* (2012). Given observation geometry (azimuth, elevation, and beam width, see Appendix A.1 for details) and model fields from the tentative analysis such as 3-D wind and precipitation mass content, observation operators compute what reflectivity or Doppler velocity would have been observed if the tentative analysis had been correct. The data assimilation system then uses these simulated observations to compute a mismatch between simulated and real observations taking into account both observation errors (generally sufficiently well estimated) and errors introduced by assumptions made in the observation operator itself (generally wrongly ignored by researchers). To design proper observation operators, it is essential to (a) correctly take into account both the beam pattern and the reflectivity field when simulating radar fields and (b) to properly account for the errors arising in simulating

Table 10.1 Measurements of any atmospheric variable per hour over a 10 × 10 km area comparable to the size of a convective cell. For this exercise, a typical surface station would be counted as reporting information for five variables (pressure, temperature, dew point, winds, and precipitation)

Data source	Number of observations
Upper air observation of any variable using planes, radiosondes, GNSS receivers, etc.	≈ 0–10 (0–8 about thermodynamics)
Surface observations of any variable	≈ 0–10 (0–4 about thermodynamics)
Geostationary satellite (per channel)	≈ 25–300
Radial velocity, assuming 60% echo coverage in a storm	≈ 15,000
Radar reflectivity	≈ 25,000

reflectivity: model fields generally describe well the precipitation mass, but an assumption must be made when converting that mass into reflectivity, not unlike the problem of estimating rainfall from reflectivity using Z–R relationships (recall Section 9.3.3). This assumption comes with errors that have a magnitude and a correlation distance that are not well known but are generally much larger than the errors on observations themselves.

Only after the data assimilation has been completed can the forecast be computed by a numerical model that simulates the expected evolution of the atmosphere. As can be appreciated, this combined data assimilation and weather forecasting process is computationally more demanding than the traditional nowcasting techniques. In controlled experiments and simulations where it is ensured that the background and its error covariance matrix are correct, its skill is superior to the latter. At the time of this writing though, this has yet to be translated into real-world benefits in the 0–2 h time frame (e.g., Fig. 10.3), primarily because of the challenges associated in obtaining accurate real-time estimates of the background and of its error structure. As a result, for the very short forecast time generally associated with severe weather, the skill of the traditional nowcasting approaches remains the reference standard. Work on data assimilation, however, remains a very active area of research, not only for trying to forecast storms but also for storm studies where one can afford to iteratively adjust the background until the time sequence of observations appears to be respected.

Data assimilation at the mesoscale and the convective scales has specific challenges compared to assimilation at larger scales. To compute a proper analysis at the global scale, measurements of a variety of quantities coming from a diversity of instruments can be used. This, together with convenient approximations such as geostrophic or hydrostatic balance, allows measurements to constrain all fields, relaxing the need for an accurate background error structure. For mesoscale or convective-scale forecasting though, radar data provide most of the new information, and they primarily provide it on the results of the storm process, for example, precipitation, not on its causes, for example, temperature, pressure, and humidity (Table 10.1). This has a variety of implications that make radar data assimilation challenging. For example, it stresses the need for accurate background error covariances that include the covariances between observed and unobserved variables. It also highlights the importance of high radar data quality and accurate observation operators, as there is essentially no other source of information that can contradict radar if its data are inaccurate. Hence, on the one hand, radar is a fantastic source of information for data assimilation and forecasting at subsynoptic scales that must absolutely be used; on the other hand, the absence of any other comparable source of information for variables unobserved by radar makes the proper data assimilation of radar data particularly difficult. We must hence explore other ways of obtaining additional information to help constrain initial conditions.

11 Additional radar measurements and retrievals

The data provided by reflectivity, radial velocity, and multiple polarization measurements are extremely valuable. However, particularly in research applications, additional information about unobserved fields or data quality would be needed to better interpret radar data, understand weather phenomena, and forecast them: What are the tangential components of the wind? How are the pressure, temperature, or humidity fields? To what extent are the radar data affected by attenuation or non-Rayleigh scattering? Many of those questions cannot be answered by radar. But some can, either by combining information from multiple radars or by processing radar data in unusual ways. In this chapter, the concepts behind a few of these approaches will be introduced.

11.1 Using multiple viewing angles

One of the key limitations of radar data is that only one of the three wind components can be measured. A solution to this problem is to combine existing measurements with those from another radar, or at least from another viewing angle. Once a second wind component is measured, the third can be derived, as will be seen at the end of this section.

Analyses of winds made by multiple Doppler radars rely on having measurements from at least two partially independent components of the wind (Fig. 11.1). Ideally, one would want to measure two horizontal wind components 90° apart to be able to simply derive the u (east–west) and v (north–south) component of the wind. In practice, this can be achieved even if two components are not 90° apart, as long as they are sufficiently different. Most researchers consider that the two components must be at least 30° apart so that measurement errors on each component do not compromise the quality of the retrieved tangential component; in that scenario, the area over which an adequate dual-Doppler wind analysis can be made forms two lobes on either side of the baseline between the two radars, as illustrated in Fig. 11.1. Increasing the baseline distance enlarges the analysis area but reduces its resolution as well as limits the ability to obtain winds at the lowest levels. For greater coverage or improved analyses, add more radars.

Two other techniques for obtaining more than one wind component are illustrated in Fig. 11.2. The first uses a rapidly moving radar, typically on an airplane, that performs conical scans along two directions such that a region of space being observed from one direction at one time will be observed from another direction a few seconds to a couple of

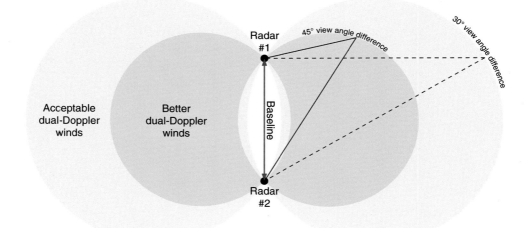

Figure 11.1. Horizontal plan view of the geometry of dual-Doppler wind analyses. More accurate winds are obtained when the viewing angle difference is the largest, resulting in radial velocity measurements that are the most independent. Given two radars, the area where acceptable (>30° viewing angle difference) and better (>45° viewing angle difference) dual-Doppler winds can be expected is shaded in light yellow and orange, respectively.

minutes later (e.g., Jorgensen *et al.* 1996). Measurements from moving radars must take into account the changing velocity and attitude (pitch, yaw, and roll) of the platform. Airborne radars such as these are used for a variety of storm studies as well as for monitoring approaching tropical storms (Lee *et al.* 2003 and references therein). Another method for obtaining the other horizontal wind component involves the use of additional (bistatic) receivers some kilometers away from the main radar (Fig. 11.2, Wurman 1994). It has the advantage over other methods that the two Doppler components from each region are measured simultaneously. Its main limitation is the large beam width of the additional receiving antennas that are necessary to get information from multiple directions for each transmitter pulse, leading to limited sensitivity, coverage, and significant sidelobe contamination (de Elía and Zawadzki 2000). Another technique for obtaining more than one Doppler component is to use interferometry approaches with spaced antennas (Zhang and Doviak 2007). Functioning best on radars that observe the same sampling volume for long periods of time and at longer wavelengths, it has found a niche in wind profilers (Cohn *et al.* 2001).

All these techniques measure components of the velocity $\mathbf{v}_t$ of targets. If these targets are precipitation, $\mathbf{v}_t$ tends to $(u, v, w - w_f)$, where (u, v, w) are, respectively, the east–west, north–south, and vertical components of the wind, and w_f is the reflectivity-weighted average terminal fall speed of hydrometeors. In most multiple Doppler configurations, it is very difficult to directly measure $w - w_f$ over large areas because the elevation angle of both radars must be very low in order to acquire measurements with significant horizontal

a) Dual-Doppler scanning strategy using airborne radars

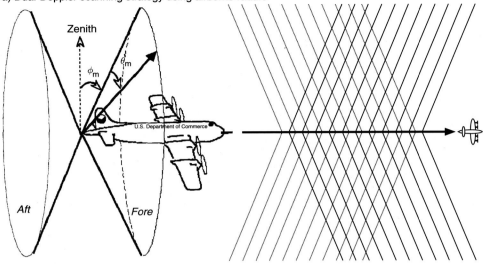

b) Multiple Doppler measurements using additional receivers

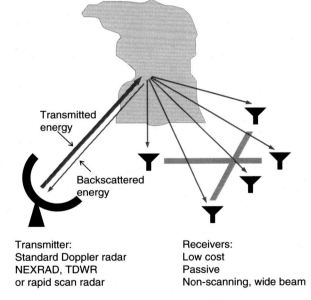

Transmitter:	Receivers:
Standard Doppler radar	Low cost
NEXRAD, TDWR	Passive
or rapid scan radar	Non-scanning, wide beam

Figure 11.2. Illustrations of two other methods for obtaining more than one wind component: a) using an airborne (or other rapidly moving) radar scanning at two angles, one slightly ahead (fore) and one slightly behind (aft) the direction of movement. The resulting scans are displayed on the right. From Jorgensen *et al.* (1996), © Copyright Springer 1996; with kind permission from Springer Science and Business Media; b) using a transmitter and multiple receivers, such as around an airport, to measure multiple wind components. Republished with permission of the AMS from Wurman (1994); permission conveyed through Copyright Clearance Center, Inc.

coverage. Perhaps most important is the fact that it is impossible to separate w from w_f using only measurements. Even if the fall speed can be estimated in rain using differential reflectivity or via a reflectivity to fall speed relationship, the errors on the vertical air velocity so obtained are comparable or exceed the magnitude of w itself for all but the most severe storms. As a result, estimations of w are often best deduced using the horizontal wind field.

When compensating for the change of air density ρ with height, the rate of change of the vertical velocity with height is due to horizontal convergence such that

$$\frac{\partial(\rho w)}{\partial z} = -\rho \left(\frac{\partial u}{\partial x} + \frac{\partial v}{\partial y} \right). \tag{11.1}$$

Therefore, given a starting value for the vertical velocity at the surface or at the top of the storm, the vertical velocity as a function of height can be obtained by integrating over height the convergence (or divergence) computed from the multiple Doppler wind analysis. The integration can be done starting from storm top or from the surface. At storm top, it is generally assumed that vertical velocity is zero. At the surface, given a surface terrain altitude $z_s(x, y)$, the vertical air velocity at the surface $w_s(x, y)$ can be estimated as

$$w_s(x,y) = u_s(x,y) \frac{\partial z_s}{\partial x} + v_s(x,y) \frac{\partial z_s}{\partial y}, \tag{11.2}$$

where $u_s(x, y)$ and $v_s(x, y)$ are the horizontal wind components at the surface. Because ρ decreases with height, uncertainties in convergence estimation lead to worse vertical velocity errors when integrating from bottom-up than from top-down. In general though, integrations from top and from bottom are combined to obtain the best possible estimate of w.

Many examples of dual-Doppler wind analyses can be found in the literature as it is a common technique used for the understanding of the dynamics of severe storms (e.g., Fig. 11.3). Software packages for their implementation have been available for some time (e.g., CEDRIC, Mohr and Miller 1983). However, wind analyses are nowadays generally being performed using a variational approach that resembles the radar data assimilation presented in Section 10.5 (e.g., Gao *et al.* 2004). For information on the latest state of the art concerning multiple Doppler wind analyses, consult the extended abstracts of presentations in the field campaign sessions of recent conferences on radar meteorology.

11.2 Multiple Doppler retrievals

The wind analysis from multiple Doppler radar measurements can be extended to retrieve unobserved quantities such as pressure and temperature fields (e.g., Sun and Crook 1997). This can be achieved by analyzing the time evolution of both the reflectivity and the wind fields. For example, Newton's law states that changes in motion are caused by forces. In the

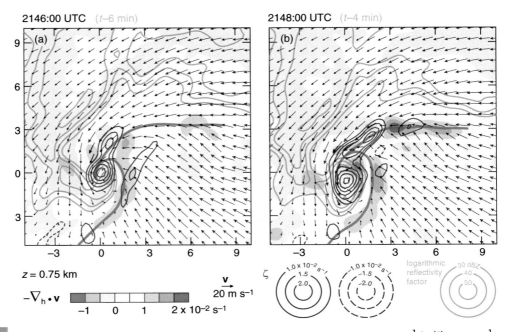

Figure 11.3. Horizontal convergence ($\nabla_h \cdot \mathbf{v}$, color shaded), vertical vorticity (ζ; red contours every 0.005 s^{-1} for $|\zeta| > 0.01$ s^{-1}; negative contours are dashed), reflectivity factor (green contours every 10 dBZ for reflectivity $\geq$ 30 dBZ), and storm-relative wind vectors ($\mathbf{v}$; plotted at every third grid point using black arrows) at 0.75 km altitude obtained from two dual-Doppler wind syntheses 2 min apart in a supercell storm just prior to tornado formation at time t. Gust fronts are indicated with heavy blue lines. On these two analyses, the area of rotation and its intensification in time can be analyzed. Axis labels are in km. Republished with permission of the AMS from Markowski *et al.* (2012); permission conveyed through Copyright Clearance Center, Inc.

atmosphere, changes in wind speeds are linked to the pressure gradient force; hence, wind accelerations can be used to retrieve pressure gradients. Used in conjunction with the hydrostatic equation and with the ideal gas law, changes in pressure gradients with height can also be linked to temperature gradients, and this opens the door to the retrieval of spatial temperature perturbations. In parallel, vertical motion generates precipitation: downdrafts lead to evaporation, while updrafts in saturated environments lead to cloud production; clouds then generate precipitation whose evolution may be observed using reflectivity. The time sequence of reflectivity can therefore provide clues on updrafts, downdrafts, and humidity. Finally, additional information on temperature can be obtained by considering heat exchanges by condensation/evaporation and possibly net heat gain by radiation. As a result, when all these interactions are considered, a lot of information can theoretically be extracted about the atmospheric state using reflectivity and wind field analyses estimated at two or more times. This process of combining the time evolution of radar information with dynamic and thermodynamic equations of the atmosphere is the basis behind retrievals.

Because these retrievals can be complex, they are increasingly done within a data assimilation framework, especially since community tools such as the WRF model and its Data Assimilation Research Testbed (DART; Anderson *et al.* 2009) are available and supported. The most difficult atmospheric quantity to retrieve by radar is humidity because while changes in temperature and pressure will cause changes to the atmospheric flow observable by radar, changes in humidity do not, except when condensation or evaporation are involved.

11.3 Near-surface refractive index

If a fixed ground target is repeatedly observed, it will be seen that the average phase measured for that target is slowly changing with time. Recalling (2.16), if the range to the target is fixed (r is constant) and none of the radar transmit and internal frequencies are changing (f is constant), then a variation in phase with time must be caused by a variation in the refractive index n along the path (Fig. 11.4). If this calculation is repeated for all fixed ground targets, fields of n, or at least of changes of n, can be obtained. This idea forms the basis for the estimation of the near-surface refractive index (Fabry *et al.* 1997). This is not to be confused with the measurement of clear echoes arising from the small-scale variability of refractive index seen in Section 2.2.2.2. Even though the refractive index is a function of pressure, temperature, and humidity as seen in (2.8), horizontal variations of refractive index mimic humidity patterns, particularly in summer conditions.

In practice, since the exact position of every ground target cannot be known to the accuracy needed for an absolute refractive index measurement (better than 1 mm), only changes in the refractive index between two times are derived by using the change in the phase measured. However, because these two times can be years apart, we may be able to choose the reference time in conditions when the refractive index field can be estimated by

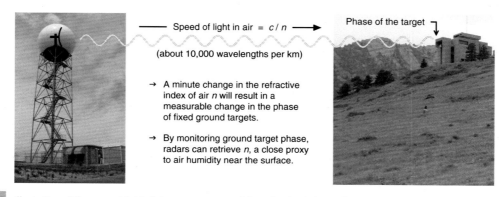

Speed of light in air = c / n

(about 10,000 wavelengths per km)

Phase of the target

→ A minute change in the refractive index of air n will result in a measurable change in the phase of fixed ground targets.

→ By monitoring ground target phase, radars can retrieve n, a close proxy to air humidity near the surface.

Figure 11.4. Illustration of the concept behind the measurement of the refractive index n of air near the surface by radar using the returns from fixed ground targets. From Roberts *et al.* (2008), © Copyright 2008 AMS.

other methods, providing a near-absolute measurement. Since the estimation of n can only be obtained along a path between a ground-based radar and ground targets, the inferred humidity measurements are only representative of atmospheric conditions very close to the surface. The curvature of the Earth and the diminishing number of ground targets with increasing range limit the range of n estimates to approximately 40 km.

Initial fears that only radars with very stable transmit frequencies could reliably make this type of measurement have been proved unwarranted (Parent du Châtelet *et al.* 2012, Nicol *et al.* 2013), and several radars using transmitters based on the less stable magnetron tube have successfully implemented refractivity measurements. However, few radars archive the phase data needed to estimate the near-surface refractive index patterns, and at this time few radar systems attempt to retrieve refractivity. One of the main reasons for this is due to the challenge of dealiasing the phase difference between the current time and the reference time: the signal is surprisingly huge, as the phase of a target 30 km away can change by more than 20,000° at S-band over the course of a year. Nevertheless, if the number of ground targets is sufficiently large, it is possible to use information over short paths to help dealias phase differences.

Despite the many difficulties associated with its proper measurement, refractive index fields provide us with a rare glimpse of the spatial patterns of humidity at the mesoscale (Fig. 11.5, supplement e11.1). Meteorologically important signatures such as convergence lines can often be detected first on refractive index measurements before fine lines can be seen on reflectivity (Weckwerth *et al.* 2005). It also provides useful constraints of temperature and humidity fields that can be used in data assimilation.

11.4 Multiple-frequency and attenuation measurements

The equivalent reflectivity factor measured by radar can sometimes be very difficult to interpret because it is affected by many processes: Is the echo dominated by small targets that obey Rayleigh scattering, or are there many larger objects that must be treated as Mie scatterers? Is a significant fraction of the echo coming from clear air reflections? And perhaps most important, how are the measurements affected by attenuation? Since a common approach for solving problems that have more than one unknown is to add more constraints, the use of data at more than one transmit frequency is a natural choice to help resolve such ambiguities.

Two somewhat distinct scenarios can be considered. At longer wavelengths, attenuation is negligible and hydrometeors ranging in size from cloud droplets to hailstones behave as Rayleigh scatterers. At those frequencies, the main ambiguity arises from whether the echo is from Rayleigh scatterers or from refractive index changes at half the scale of the wavelength (Section 2.2.2.2, Fig. 2.4). But how reflectivity varies with wavelength is different for both types of targets: for Rayleigh scatterers, it does not change significantly with the radar wavelength λ, because the reflectivity factor (3.1) and the radar equations (3.2) were designed to obtain frequency-independent results for Rayleigh scatterers; for

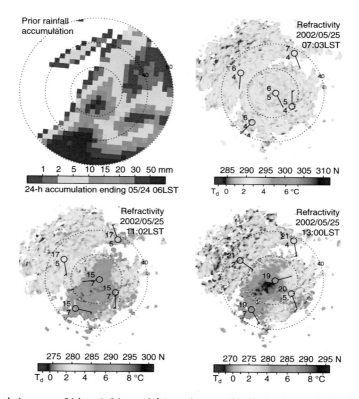

Figure 11.5. Rainfall accumulation over a 24-h period (upper-left image) responsible for the time evolution of refractivity ($N = 10^6(n-1)$, shown in the following images) and the corresponding dew point temperatures (T_d) on the first sunny morning following the rain. Thanks to very weak winds and the uneven rain in the previous days, we observe the gradual appearance of regions of different humidity, with more humid regions corresponding to areas of heavy rain on the previous days. Range rings are every 20 km. From Fabry (2006), © Copyright 2006 AMS.

echoes from refractive index variations, also known as Bragg scatterers, reflectivity varies as $\lambda^{11/3}$, corresponding to an increase of 11 dB for each doubling of wavelength. Rayleigh and Bragg scatterers can be easily distinguished from measurements made with two distinct frequencies (Fig. 4.14, see also Knight and Miller 1998). The strongest Bragg scatterers originate from cloud tops, inversion layers, boundary layer inhomogeneities, and occasionally from fires and volcanic eruptions.

The second situation where multiple wavelength measurements can provide additional insights is at shorter wavelengths, where the interpretation of reflectivity measurements from hydrometeors can be complicated by both non-Rayleigh scattering and attenuation that unfortunately often occur together. At very short ranges, and in the absence of liquid precipitation, attenuation can often be neglected, the difference between the reflectivity at both wavelengths being primarily a function of the size and shape of the crystal or snowflake being observed. However, as the path length increases, especially in the presence of liquid precipitation, attenuation becomes the main cause of the reflectivity difference between the two radar wavelengths. Attenuation is an interesting quantity to estimate: at shorter wavelengths (≤ 1 cm), away from precipitation, it varies mostly as a function of humidity, and its measurement therefore can provide humidity information (Ellis and Vivekanandan 2010). At somewhat longer wavelengths (1–5 cm), attenuation in precipitation can actually be used to measure rainfall (Atlas and Ulbrich 1977; Ryzhkov *et al.* 2014).

An interesting approach has been proposed by Thompson *et al.* (2011) for estimating radome and precipitation attenuation on operational radars. It is based on the concept that a good absorber is an efficient emitter, and that on azimuths where attenuating precipitation is considerable, a greater level of noise is expected compared with other azimuths because of the microwave emissions from that precipitation. Therefore, by accurately measuring the strength of the noise, it is possible to estimate how much path-integrated attenuation occurs in that azimuth.

These last two ideas, combining the measurements from radars with different wavelengths and measuring the emissions coming from precipitation, also form the basis of many systems designed to study clouds and precipitation from space.

Cloud and spaceborne radars

There is a new role for radar since the eve of the twenty-first century. It arose from the growing interest in clouds and global precipitation, partly fueled by the wish to better understand the Earth energy balance and hydrological cycle, and partly in response for the need to improve the simulation of clouds and of precipitation in climate models. Both of these needs require shorter wavelength radars: ground-based cloud studies are best made with radars that have narrower beam widths and greater sensitivity to cloud droplets and ice crystals, while global cloud and precipitation monitoring requires radars that are both small enough to be carried on satellites, yet have sufficient resolution and sensitivity to make useful measurements. These radars also tend to be used differently than weather radars, the focus being more on cloud and precipitation microphysics and the characterization of long-term statistics rather than on storm processes and minute-to-minute weather surveillance and nowcasting. These differences in focus have led to the emergence of two distinct communities of radar meteorologists who largely work in isolation of one another. The use of shorter wavelengths either from a ground-based or a spaceborne platform gives rise to a combination of challenges and opportunities that are unique to these radars and deserve a special discussion in this chapter.

12.1 Cloud radars

12.1.1 Radars for cloud studies

The contribution of clouds on the Earth energy balance, and how it may change as a result of man-made perturbations such as greenhouse gas and aerosol emissions, remains one of the greatest sources of uncertainties in climate simulations and predictions. There is hence a societal need to better characterize clouds. In parallel, process studies of the formation of clouds and how precipitation initiates remain an active research area. Both of these research interests called for the use of new data.

But the measurements required are particularly difficult to make. In the context of studies of the radiation budget of the planet, even an extremely thin liquid cloud made of very small cloud droplets noticeably changes the visible and infrared radiation and should therefore be quantified. A back of the envelope calculation reveals that a thin liquid cloud with 5 μm drops and a liquid water content of 0.015 g/m^3 should have a reflectivity of approximately -55 dBZ, many orders of magnitude smaller than the echoes observed by traditional weather radars (Table 3.2). More than for any other use of radar, it becomes essential to maximize the power returned from such weak targets:

$$P_r = \underbrace{\frac{1.22^2 0.55^2 10^{-18} \pi^7 c}{1024 \log_e(2)}}_{\text{Constants}} \underbrace{\frac{P_t \tau D_a^2}{\lambda^4}}_{\substack{\text{Radar} \\ \text{parameters}}} \underbrace{\frac{T(0,r)^2}{r^2}}_{\text{Path}} \underbrace{|K^2| Z}_{\substack{\text{Target} \\ \text{properties}}}, \qquad (12.1)$$

Most clouds being small compared with precipitation systems, high-resolution measurements are preferred, also implying shorter transmit pulses τ and smaller beam widths achieved by maximizing D_a/λ (2.11). These combined requirements have pushed radars designed for cloud studies toward higher frequencies or shorter wavelengths, such as Ka-band (35 GHz or 8.5 mm) and W-band (94 GHz or 3.2 mm), that currently maximize our ability to make cloud measurements (Kollias *et al.* 2007). This is why these millimetric radars are now referred to as cloud radars. Figure 12.1 shows an example of data collected by cloud radars, illustrating what can be observed thanks to their higher resolution and sensitivity.

Figure 12.2 illustrates in one plot many of the advantages and challenges associated with moving to shorter wavelengths. As wavelength decreases, the sensitivity of the radar improves, and the risk of having weak echoes from variations in the refractive index of air diminishes. This is particularly important for cumulus clouds whose echo at longer wavelengths is dominated by reflections from the variability of air temperature and humidity (Fig. 4.13), masking the returns from the cloud droplets themselves. One of the consequences of moving to shorter wavelengths is that precipitation-size hydrometeors increasingly become non-Rayleigh scatterers, their diameter becoming comparable to the size of the wavelength (Figs. 2.2 and 2.4). The result is that the echo strength from

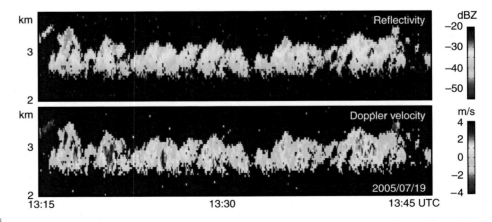

Figure 12.1. Time–height section of reflectivity (top) and Doppler velocity (bottom) of cumulus clouds observed by a W-band cloud radar at the Atmospheric Radiation Measurement Southern Great Plains site in Oklahoma, USA. Note the increase of reflectivity with height for each cumulus as the droplets lifted by updrafts grow larger. Image courtesy of Pavlos Kollias.

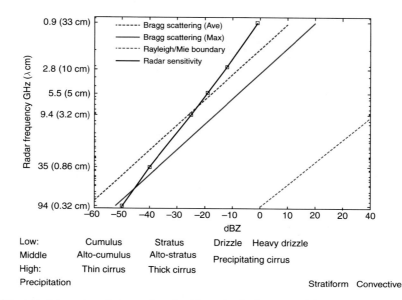

Figure 12.2. Plots of the typical dependence between the reflectivity and radar frequency of the sensitivity of several existing radar systems (black line), the expected average and peak intensity of echoes arising from the variability of the refractive index of air (respectively the solid and dashed red lines), and the reflectivity of weather echoes beyond which significant non-Rayleigh scattering is expected (blue dashed line). On the reflectivity axis, the expected Rayleigh reflectivity of different cloud and precipitation targets is indicated. For a given radar frequency or wavelength, the range of reflectivity values between the red and the blue lines correspond to targets that obey Rayleigh scattering and are hence easier to interpret for atmospheric studies. Republished with permission of the AMS from Kollias *et al.* (2007); permission conveyed through Copyright Clearance Center, Inc.

the largest hydrometeors is considerably reduced compared with what is expected for Rayleigh scatterers, leading to a breakdown of the traditional assumption that Z is proportional to D^6. This reduction is in addition to the one caused by attenuation that also becomes important at shorter wavelengths. Furthermore, because the larger drops affected by non-Rayleigh scattering fall faster than the unaffected smaller droplets, the measured fall speed will also be different than at longer wavelengths. The net result is that, in precipitation, it is more challenging to quantitatively interpret measurements from millimeter radars compared with those obtained at longer wavelength. For combined studies of clouds and precipitation, it then becomes advantageous to use multiple radars with different wavelengths as they provide complementary information.

The example in Fig. 12.3 illustrates many of the differences between echoes from cloud and from precipitation radars. In this type of event, the difference in sensitivity between the two systems does not lead to a large change in echo coverage, though echoes are detected slightly higher with the W-band than with the X-band radar. More obvious to notice are the effects of attenuation and of non-Rayleigh scattering in precipitation. Attenuation is most visible in rain, as the echo strength steadily decreases with height at W-band, with swaths of

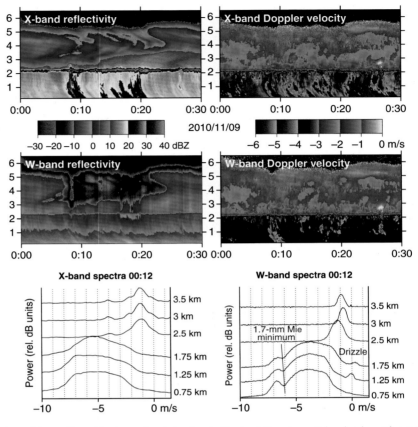

Figure 12.3. Comparison of data collected by two collocated radars in Montreal, Canada: an X-band radar with a 1.8° beam and a W-band with a 0.25° beam. At the top, time–height sections of reflectivity and Doppler velocity measured at both wavelengths are contrasted. At the bottom, the Doppler spectra measured at 00:12 by both radars are shown for a few altitudes. Note how the W-band spectra are narrower than those from the X-bands in rain and in snow, leading to the detection of a weak drizzle mode. Also notice how the power corresponding to 1.7-mm diameter raindrops is notched.

much attenuated signals being observed when precipitation is the strongest at X-band. The time variations of path-integrated attenuation also lead to vertical striations in reflectivity in snow. Non-Rayleigh scattering at W-band manifests itself in at least four ways: (a) the reflectivity of rain at very low levels is already different between the two radars even before attenuation could have a significant effect; (b) the bright band, made of large wet hydrometeors, has essentially disappeared at W-band; (c) the Doppler velocity in rain, and to a much lesser extent in snow, is reduced as a result of the relative loss of sensitivity to larger and faster falling hydrometeors; and (d) in spectra, a relative minimum in power returned can be observed for raindrops of 1.7 mm diameter. This last effect will be discussed further in the next section. Last but not least, Doppler spectra measured by narrow-beam radars such as cloud radars are narrower than that of wider-beam systems, giving them a greater ability to resolve multiple-peaked spectra. In this example, a small secondary mode of drizzle can be better observed by the W-band system than by the X-band radar. Each of these differences represents an opportunity when combining the information from radars with different frequencies.

Even at millimetric wavelengths, radars do miss small liquid clouds and very thin ice clouds. As a result, probably more than in any other radar research, these radars are paired with other instruments such as lidars and microwave radiometers that complement their measurements. Lidars, or "light radars," are laser-based remote sensors that operate at near-visible or at infrared wavelengths (λ generally between 0.3 and 10 μm). The atmospheric properties that they detect are therefore much closer to what influences radiation processes. But, like our eyes, they lack the ability to make observations through optically thick clouds. Microwave radiometers measure the emissions of microwaves originating from oxygen, water vapor, clouds, and precipitation, and use measurements at different frequencies to distinguish the contributions from different emitters. Functioning at wavelengths comparable to that of cloud radars, they can be used to quantify cloud amounts. But because they are passive remote sensors, they cannot make high-resolution measurements in range, and their data are often limited to path-integrated quantities. Hence, none of these instruments provide an ideal set of measurements; but their data can be combined to take advantage of the respective strengths and benefit from their synergy as shown in Fig. 12.4. The outcome is a more complete cloud coverage statistics or the successful retrieval of additional information on cloud properties (e.g., Frisch *et al.* 1995, Hogan *et al.* 2001).

For a variety of cloud studies applications, the scanning strategy of choice is to point the radar vertically. This simple scanning strategy permits the highest sensitivity and resolution by minimizing the range to the cloud as well as increasing the time of observation of any sampling volume while limiting maintenance needs. It also allows the cloud radar to obtain one of its most precious measurements: Doppler spectra.

12.1.2 Doppler spectra from cloud radars

Doppler spectra provide the distribution of Doppler velocities of targets within a given sampling volume. At vertical incidence, this corresponds to the distribution of fall speed of targets (Fig. 5.3). Doppler spectra are measured and archived by a variety of radars such as longer wavelength wind profilers and vertically pointing weather radars providing insights on storm dynamics, such as updrafts and downdrafts, as well as on microphysics. Initial hopes of using Doppler spectra to estimate both updrafts and the drop size distribution of rain were frustrated by a variety of factors complicating the interpretation of spectra: at weather radar wavelengths, it is difficult to take into account the effect of the mean vertical air velocity; at profiler wavelengths, separating clear air echoes from the echoes of very light precipitation proved extremely challenging; and at all wavelengths, broadening of the measured distribution of Doppler velocities by a variety of phenomena (Fig. 5.2) led to a smoothing or a washout of spectra, where narrow nearby peaks in spectra broaden and merge to become one, frustrating many efforts to study cloud and precipitation microphysics.

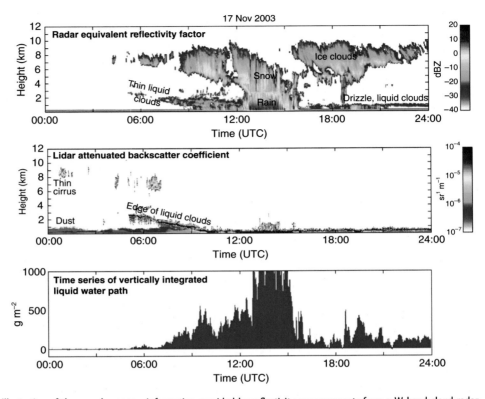

Figure 12.4. Illustration of the complementary information provided by reflectivity measurements from a W-band cloud radar (top), backscatter coefficient (or signal strength) uncorrected for attenuation from a lidar (center), and retrieved vertically integrated liquid water content from a microwave radiometer (bottom) measured over a 24-h period at Chilbolton, UK. Republished with permission of the AMS from Illingworth *et al.* (2007); permission conveyed through Copyright Clearance Center, Inc.

At least two factors favor the use of spectra from cloud radars for atmospheric studies, especially at W-band. First, the narrower beam widths of cloud radars reduce the broadening of Doppler spectra. This reduction is not sufficient so as to ignore the problem, but leads to an increase in the sharpness of spectra, making it possible for example to start distinguishing cloud droplets from the smallest precipitating drops (Luke and Kollias 2013). Second, under specific circumstances (when rainfall contains some large drops but is not yet too intense to completely attenuate the radar signal), W-band radar spectra of rainfall can be used to retrieve vertical air velocity. This unexpected result arises as a consequence of Mie scattering. For some specific drop sizes, the power returned is much reduced. At X-band (3.2-cm wavelength), the first of these reductions would be observed for drops with a 17-mm diameter (Fig. 2.2), if they commonly occurred in nature. At W-band (3.2-mm wavelength), the same phenomenon is seen for 1.7-mm drops (Fig. 12.3). Since the fall speed of 1.7-mm drops can be computed with altitude given air density, the difference between the measured and expected fall speed of these drops must be due to air velocity (Lhermitte 1988). Though the detection and interpretation of the Mie minimum can be complicated by turbulence, this estimation of air velocity is a useful bonus for a variety of microphysical and dynamical studies (Giangrande *et al.* 2012).

Cloud radars are used for microphysical research in several countries, and their use is slowly expanding to support climate-related research. Their data, and that of other sensors, are also precious for validating measurements from other radars designed to study global climate: spaceborne radars.

12.2 Spaceborne radars

A large fraction of land areas and most oceanic regions are not observed by ground-based radars. To obtain a truly global perspective, a radar on a spaceborne platform is therefore a necessity. Since the late 1990s, meteorological radars have been making observations from space, enhancing our understanding of precipitation and clouds, especially away from regions covered by ground-based radars.

A priori, there should be no significant differences between radars in space and other radars. In practice, spaceborne radars must be light and compact at launch to be carried on orbit, and they must not consume too much electricity. On low orbit, the radar speed is close to 8 km/s with respect to the surface, which limits dwell time and complicates Doppler measurements. And on geostationary orbit, the 36,000-km range to the surface imposes other challenges, particularly on spatial resolution. These high demands on sensitivity and compactness have hence pushed spaceborne radars toward shorter wavelengths in a manner that similar constraints did for cloud radars.

12.2.1 The satellite platform and its peculiarity

There are a few particular features to radar measurements from space that arise from the different measurement geometry and from orbital mechanics as illustrated in Fig. 12.5.

Even on low-Earth orbit, the far range of the radar from the intended targets in the troposphere just above the ground has a variety of consequences. Despite a narrow beam width (0.7° on the illustration), the area illuminated by the radar is fairly large, of the order of 5 km in diameter. And if the radar scans away from the vertical, the pulse at any given range will have a different altitude on one side of the beam compared with the other (note the axis of the small transmit pulse line within the troposphere for the three beams illustrated in Fig. 12.5). This will also extend in height the contamination from ground targets. Another consequence is that the range interval over which targets are expected in the troposphere is small, both in absolute terms and with respect to the total range to the satellite. This geometry minimizes the impact of attenuation, allows the radar to transmit

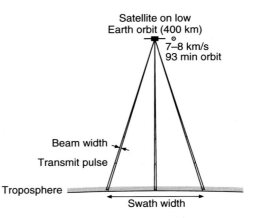

Figure 12.5. Illustration to scale (except for the size of the satellite) of the measurement geometry of a spaceborne radar on low-Earth orbit (400-km altitude). In this example inspired from the characteristics of the Ku-band radar on the core satellite of the Global Precipitation Measurement (GPM) Mission, three beams covering the two extremities and the center of the swath are illustrated, even though more are being used in between. The satellite is assumed here to orbit out of the page.

many pulses before it receives echoes, and results in a sensitivity that varies little with range (or height). Hence, for example, the stated sensitivity of the GPM precipitation radar (Ku-band) is approximately 17 dBZ.

Orbital mechanics also influence radar measurement strategies. If a radar scans off nadir, it must do so very rapidly as a satellite on a low-Earth orbit moves at close to 8 km/s with respect to the surface. This limits both the averaging (or dwell) time over any single beam and the number of angles that can be scanned. The number of angles scanned constrains both the swath width over which measurements are taken and the resolution of measurements within the swath. Next, because the orbital time at such altitudes is slightly above 90 min, the next line or swath of measurements can be as far as 2500 km from the previous one as a result of the rotation of the Earth under polar-orbiting satellites, while swath widths are at most of the order of a few hundred kilometers. Therefore, depending on latitude, there could be many days between revisits at the same location. This limitation makes the use of radars on low-Earth orbit for operational weather surveillance impractical, but is one that climate-oriented researchers have learned to accept. Finally, many low-orbiting satellites are on what are called sun-synchronous orbits, meaning that all their overpasses occur at the same solar time. CloudSat and the other satellites of the A-Train, for example, are on a 1:30 orbit, and take all their measurements around 01:30 and 13:30 solar time as a result of the mechanics of the orbit chosen. The use of sun-synchronous orbits is particularly appropriate for sensors relying on visible light to guarantee good lighting on half the overpasses, but set limits on the type and value of climatological statistics that can be made with the datasets collected. That is why satellites where radar is the main sensor tend not to be in sun-synchronous orbits such as TRMM and the GPM core satellites.

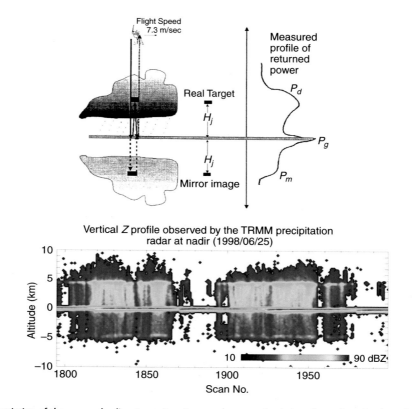

Figure 12.6. Top: Description of the process leading to a mirror image echo appearing below the surface. On the right, the measured profile of returned power from the direct echo (P_d), the ground or sea surface (P_g), and the mirror image (P_m) is sketched. Republished with permission of the AMS from Li and Nakamura (2002); permission conveyed through Copyright Clearance Center, Inc. Bottom: Vertical profile of reflectivity measured by the TRMM precipitation radar at NADIR on June 25, 1998, illustrating the direct, sea surface and mirror image echoes. Image courtesy of Kenji Nakamura.

However, observations from space offer unexpected advantages for the quantitative interpretation of radar, especially for attenuation correction. The first arises from the formation of a mirror image of echoes caused by the partial reflection of radar waves by the sea surface (Fig. 12.6). Two images of the same reflectivity pattern can therefore be obtained, one direct, and one after reflection by the surface. The complementary bottom-up and top-down views can then be compared to estimate the strength of radar attenuation as a function of altitude. This approach is known as the mirror image technique (Li and Nakamura 2002). The second approach, called the surface reference technique (Meneghini *et al.* 2000), relies on the difference between the expected strength of the echo from the surface and that measured by the radar. This technique therefore can also be used to provide an estimate of the path-integrated attenuation caused by precipitation.

12.2.2 Context behind spaceborne radars

Spaceborne radars are still in their infancy and are primarily driven by the need to better characterize global precipitation and cloud characteristics. At this time, spaceborne radars are either focused on precipitation measurements such as those on GPM, or on cloud measurements such as those on CloudSat and EarthCARE. Precipitation measurements for climate studies are difficult to make: the wavelength(s) must be chosen to avoid biases due to attenuation and minimize the effects of varying Z–R relationships, yet the system must be capable to quantify both the weaker precipitation of polar regions and the strongest convective cores of tropical areas. For climate needs more than for any other investigations, biases in estimates cannot be tolerated, especially if they vary with time or location. These constraints have led to the development of dual-wavelength spaceborne radars such as the one on board the GPM (Fig. 12.7) where the information from both wavelengths can be combined with the mirror image and surface reference techniques to better estimate attenuation as well as precipitation intensity. For cloud studies, achieving the needed sensitivity is a critical design parameter. As a result, spaceborne cloud radars are highly sensitive W-band systems that at this time do not scan.

To help fulfill their science missions, spaceborne radars are often teamed with microwave radiometers (for precipitation-focused missions) and lidars (for cloud-oriented missions). These complement, extend, and help in the understanding of spaceborne radar measurements as is the case with ground-based cloud radars. The strengths of radars within this combination of sensors remain their ability to make measurements at high-range resolution into clouds or precipitation and to allow the study of their vertical and horizontal structure over the globe (Figs. 12.8 and 12.9).

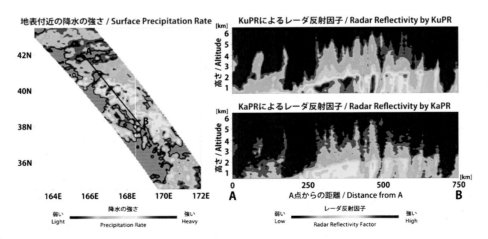

Figure 12.7. Imagery of the first data collected by the precipitation radars of the core satellite of the GPM. Left: Derived surface precipitation rate. The A-B line shows the location where the vertical sections on the right have been made. Right: Vertical section of reflectivity on the Ku-band ($\lambda = 2.2$ cm, above) and the Ka-band ($\lambda = 8.5$ mm, below) radars. Image courtesy of NASA and JAXA.

Precipitation frequency derived by the TRMM radar, 1998–2008

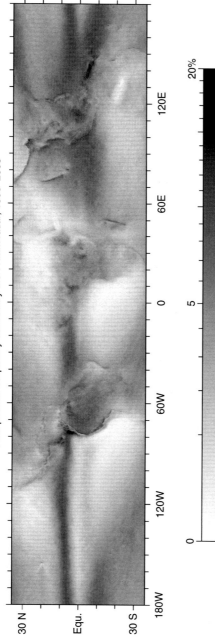

Figure 12.8.

Frequency of precipitation detected by the TRMM Precipitation Radar for the period between January 1998 and December 2008 using the data from Nesbitt and Anders (2009). On this image without geographical overlay, the influence of continents, islands, and mountain chains on precipitation is clearly revealed. The Intertropical Convergence Zone (ITCZ) is also very well delineated.

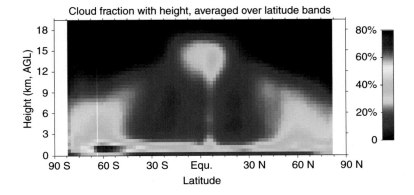

Figure 12.9. Zonal annual mean vertical cloud fraction derived from the CloudSat radar and CALIPSO lidar from 2006 to 2011. Image courtesy of Jennifer Kay (University of Colorado) using methods described in Kay and Gettelman (2009).

Challenges remain. There is ample evidence that the sensitivity of spaceborne radars is not yet sufficient to detect weak precipitation and thin clouds such as those found in polar regions, as most snow events have reflectivity weaker than the detection threshold of the GPM radar. In parallel, combating attenuation and converting reflectivity and Doppler velocity measurements into reliable long-term precipitation measurements for climate research still pose problems. This does and should not stop researchers from dreaming though. Some already imagine the possibility of putting weather radars on geostationary orbit at 36,000 km altitude to better monitor severe weather, especially over oceans.

13 What does radar really measure?

Until this point, the radar measurement process has been treated as a black box, introducing in the early chapters only what was necessary to allow a nonresearch user to understand radar data and how they can be corrupted. But the capabilities of the instruments we use dictate what can be measured, and therefore influence our perception of reality. To properly comprehend what we observe, a good knowledge of the capabilities of the instrument we use is hence a prerequisite. For radar in particular, the dependence between what can be observed and the characteristics of the instrument is often complicated, being dictated by the complex relationships between the theoretical and engineering aspects of the measurement process together with the properties of the fields being observed. To answer the question "what does radar really measure" requires a deeper understanding of how the engineering, physical, and meteorological considerations are intertwined. This understanding is essential for a proper quantitative interpretation of observations.

In this chapter, the measurement process is reexamined by first studying it from the physics and engineering perspectives, and then completing the picture by adding meteorological considerations.

13.1 The radar system

Figure 13.1 and the electronic supplement e02.1 illustrate the basic elements of a radar system. In order to better understand them, we shall follow the many steps in the travels of a radar pulse, first focusing on the radar hardware.

13.1.1 From the transmitter to the antenna

The travels of radar waves start at the transmitter. For most weather radars, the transmit signal is generally a short high-power pulse shaped by a modulator or a waveform generator, although there are radars that use a continuous wave signal (Skolnik 2008). Transmitters can be subdivided into two families. In the first one, the radar pulse is generated by a high-power oscillator that is fed a strong pulse of current and emits a strong microwave pulse in exchange. The most common oscillator is the magnetron, which powers many radars and most microwave ovens. In the second type of transmitter, a short pulse is generated by low-power frequency sources and amplified by specialized amplifiers such as

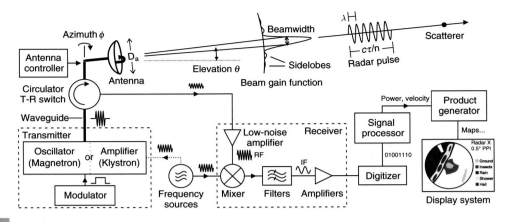

Figure 13.1. Block diagram of a monostatic (single-antenna) weather radar.

klystron tubes. Historically, the difference was important as only amplifier-based radars could be used to get Doppler information, but this is no longer the case. Nevertheless, amplifier-based radars still make better phase measurements and provide more flexibility such as intentional phase coding (Sachidananda and Zrnić 1999; Frush *et al.* 2002). Pulse radars normally transmit microwaves at a single frequency that dictates the wavelength of the radar wave in the atmosphere. Depending on the type of transmitters or weather radars, transmit pulses can reach a peak power of a few hundred watts to a few mega-watts. These pulses have typically a very short duration τ, of the order of 1 μs. The duration of the pulse determines the pulse length $c\tau/n$ that illuminates the targets. Furthermore, unless pulse compression or processing techniques are used, the resolution of the radar in distance, or range, is determined by the pulse length. After a pulse is fired, the transmitter becomes silent to allow the receiver part of the radar to capture the weak echoes scattered back to the radar.

Before that happens, the pulse travels toward the antenna in a hollow metal tubing with a rectangular section known as a waveguide. On its way, it passes by the circulator. The circulator acts as a traffic cop, directing transmitter power toward the antenna and the soon-to-come echoes toward the receiver circuitry, while a transmit-receive (T-R) switch further insures that none of the high transmit power gets into the ultra-sensitive receiver.

13.1.2 Antenna and beam shaping

The role of the antenna is to serve as an interface between the radar system and the atmosphere. In weather radars, we want to focus as much of the energy as possible in one specific direction determined by the azimuthal angle ϕ clockwise from north and the elevation θ above the horizon. The direction pointed by the radar generally changes with time under the command of the antenna controller.

Figure 13.2 graphically illustrates the principle of operation of a reflector antenna. Briefly, the radar wave that traveled in the waveguide is then directed toward a reflector

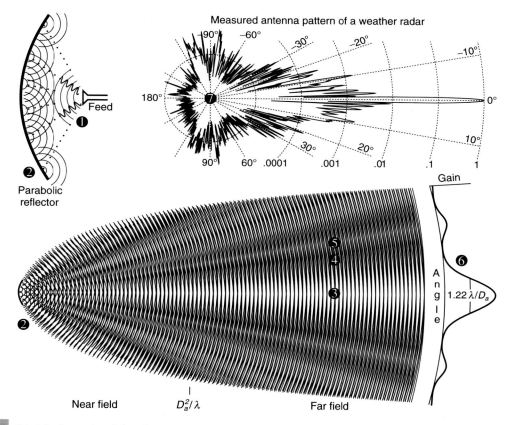

Figure 13.2. Principle of operation of the reflector antenna. At the end of the waveguide, a feed directs the radar wave toward a reflector ①. Concentric rings indicate the location in space where the phase of the wave reaches a specific value, here $\pi/2$. When the wave arrives on the dish-shaped reflector ②, it is scattered in all directions. But beyond a sufficient distance from the antenna (in the far field), a pattern emerges: along the bearing pointed by the antenna, waves reflected by different sections of the reflector are all in phase and interfere constructively ③. This maximizes the power emitted toward and received from that direction. In other directions, the waves reflected by different sections of the reflector are out of phase and interfere destructively, either perfectly as in ④, or partially as in ⑤. The resulting beam pattern ⑥ is a main lobe with a half-power beamwidth of approximately 1.22 λ/D radians in the direction pointed by the antenna, and diminishing sidelobes as we move away from the main lobe. Even if most of the energy is focused on the desired bearing, some of it is emitted and received from sidelobes pointing in all directions. For example, on the top right of the figure, the relative amount of power emitted in a given direction when the radar transmits for the McGill University S-band radar is shown ⑦; this is also the relative sensitivity of the antenna system on reception.

by a feed. If you want to observe such a setup from up close, simply go to a nearby electronics store selling satellite TV systems and look at the reception antenna on display. The radar wave is scattered by the reflector. The geometry of the system, that is, the position of the feed and the shape of the reflector, ensures that as much of the radiation as possible is focused in one direction. In systems operating at long wavelengths ($\lambda \gg 10$ cm), building large reflectors becomes impractical. Instead of having a single reflector that reflects the radar waves at the appropriate phase in order to obtain the interference pattern observed in Fig. 13.2, an array of antennas is used to achieve the same result (supplement e13.1).

The specifics of how well a given antenna is able to focus the energy in one direction depend on wavelength, the design of the feed and of the reflector, and that of potential obstacles to the emitted radiation such as the radome that protects both the antenna and the pedestal system from the wind and rain. As a result, it is impossible to write an accurate equation to represent the antenna pattern with its sidelobes. For applications where the representation of sidelobes can be ignored, a common approximation is to assume that the antenna pattern follows a Gaussian distribution, the gain (or directivity) $G(\delta\phi,\delta\theta)$ of the antenna being, for low enough elevations ($\theta < \pi/2 - 5\theta_{\text{beam}}$)

$$G(\delta\phi,\delta\theta) \approx \frac{4\log_e(2)\cos\theta}{\pi\phi_{\text{beam}}\theta_{\text{beam}}}\exp\left\{-4\log_e(2)\left[\frac{(\delta\phi\cos\theta)^2}{\phi_{\text{beam}}^2}+\frac{\delta\theta^2}{\theta_{\text{beam}}^2}\right]\right\}, \qquad (13.1)$$

where $\delta\phi$ and $\delta\theta$ are the azimuth and elevation angle deviations with respect to the beam axis, and ϕ_{beam} and θ_{beam} are the half-power beamwidths of the antenna in azimuth and elevation. In (13.1), the $\cos\theta$ term allows for a simple representation of the fact that, as elevation increases, lines of constant azimuths get closer together. For typical meteorological antenna designs, $\phi_{\text{beam}} \approx \theta_{\text{beam}} \approx 1.22\lambda/D_a$, D_a being the diameter of the reflector. Equation (13.1) expresses the one-way directivity of the antenna. For radar applications, the antenna is used twice, first on transmit to focus the energy in one direction, and then on receive to capture the signals arriving from that same direction. The net angular weighting function that describes the sensitivity of the radar to targets in different directions is hence proportional to G^2 ($\approx\pi\phi_{\text{beam}}\theta_{\text{beam}}G^2(\delta\phi,\delta\theta)/(2\log_e(2)\cos\theta)$). The radar therefore receives four times more power from a target on the beam axis than from a similar target $\theta_{\text{beam}}/2$ away from its axis. However, because the reflectivity of weather targets can span several orders of magnitude (several tens of dB), if the antenna points in a region with very weak targets, a strong target $2\theta_{\text{beam}}$ away can still dominate the signal received. As a result, for example, the echo top height of strong storms often tends to be overestimated by radar, and returns from strong ground echoes are received over a wider angular sector than their actual physical dimension.

If sidelobes are to be included, then G becomes very complicated and there is typically no analytical function that can represent it. A simple and perhaps simplistic approximation is to assume that the beam pattern can be reproduced by the sum of a Gaussian function representing the main lobe pattern and of an exponential function representing the envelope of the sidelobe pattern:

$$G(\delta\phi,\delta\theta) \approx \frac{4\log_e(2)(1-f_{\text{lobes}})\cos\theta}{\pi\phi_{\text{beam}}\theta_{\text{beam}}}\exp\left\{-4\log_e(2)\left[\frac{(\delta\phi\cos\theta)^2}{\phi_{\text{beam}}^2}+\frac{\delta\theta^2}{\theta_{\text{beam}}^2}\right]\right\}$$

$$+\frac{0.306\cos\theta}{\phi_{\text{lobes}}\theta_{\text{lobes}}}f_{\text{lobes}}\exp\left[-2\log_e(2)\sqrt{\frac{(\delta\phi\cos\theta)^2}{\phi_{\text{lobes}}^2}+\frac{\delta\theta^2}{\theta_{\text{lobes}}^2}}\right], \qquad (13.2)$$

where f_{lobes} is the fraction of the integral of the gain belonging to the sidelobes, and ϕ_{lobes} and θ_{lobes} are the half-power widths of the sidelobe envelope in azimuth and elevation, respectively. For typical scanning radars, ϕ_{lobes} and θ_{lobes} correspond to about 10°, and f_{lobes} is of the order of 0.25, leading to a sidelobe term which is approximately 30 dB below the mainlobe term at $\delta\phi = \delta\theta = 0$.

13.1.3 The received signal

If some of the radar wave is scattered by targets in all three dimensions, then a tiny fraction of the transmitted energy is necessarily reflected back to the radar. This very weak signal is focused by the reflector back into the feed and the waveguide. It then reaches the circulator that redirects it toward the receiver circuitry (Fig. 13.1).

While the transmitted power is often close to a megawatt for weather surveillance radars, the received power ranges from a few milliwatts to as little as 10^{-15} W, that is, up to 20 orders of magnitude (!) smaller than the transmit pulse. The first task of the receiver is hence to amplify this extremely weak signal. Once adequately amplified, the received signal at microwave frequency centered on f is mixed with a reference wave from an internal local oscillator at a nearby frequency f_{LO} to obtain, among other things, a signal with a lower intermediate frequency (IF) centered on $f - f_{LO}$ that can be first selected by filters, and then more easily amplified and digitized than the original return could have been. If your radar was manufactured after 2025, there is a good chance that the receivers of your era do not need the mixing step anymore, digitizers having likely become fast enough to work directly on the received signal at microwave frequencies.

As seen previously, the amount of signal power received at the antenna is determined by (3.2). In parallel, noise is being received from atmospheric emissions originating along the path s pointed by the radar:

$$P_N = kBT_A \approx kB\left[T_{co}T(0,\infty) + \int_0^\infty \alpha(s)T(s)T(0,s)ds \right], \qquad (13.3)$$

where P_N is the noise power, k is the Boltzmann constant (1.381×10^{-23} J/K), B is the bandwidth of the receiver (in Hz), and T_A is called the noise temperature of the antenna. As shown in (13.3), T_A is a function of what is left of the noise temperature of the cosmic microwave background T_{co} (2.7 K) once it went through the atmosphere with a transmittance $T(0,\infty)$, to which is added the noise from the atmosphere. That atmospheric noise is due to the integral along the path of the specific emissivity (and absorptivity) α at location s multiplied by the local temperature $T(s)$ and modified by the transmittance $T(0, s)$ between that location and the radar. Note that the transmittance $T(0, s)$ is linked to the absorptivity via

$$T(0,s) = exp\left[-\int_0^s \alpha(s')ds' \right]. \qquad (13.4)$$

Equations (13.3) and (13.4) are traditional radiative transfer equations for which scattering is ignored, whereby at each segment ds along the path, the energy headed to the radar is reduced to $(1-\alpha(s)ds)$ of its original intensity by attenuation, and a new contribution proportional to $(\alpha(s)T(s)\,ds)$ is added in its place. Both the signal power P_r and the noise power P_N are measured simultaneously by the radar, noise acting as a "competition" to the signal power we want to measure. Using (13.3), it can be shown that the antenna temperature T_A is generally bound between two numbers: at one end, a perfectly transparent atmosphere ($\alpha(s) = 0$ and $T(0, s) = 1$ everywhere) would yield a T_A of T_{co}; at the other end, an opaque atmosphere (high $\alpha(s)$ for small s and $T(0, s) \approx 0$ beyond), T_A approaches the environmental temperature T, generally of the order of 290 K. In the latter case, for a receiver bandwidth of 1 MHz, P_N reaches 4×10^{-15} W. The amount of noise originating from the atmosphere sets the antenna temperature T_A somewhere between these two numbers, a thin transparent atmosphere such as what is measured at high elevations and at low frequencies yielding the smallest noise power, while lower elevation measurements where path attenuation becomes important are affected by stronger noise.

But the story of noise does not end here (supplement e13.2). A similar radiative transfer process occurs inside the radar receiver itself: every component from the waveguides to the digitizer has some fractional loss. As a result, all along the receiver chain, noise power increases relative to signal power, noise increasing by anywhere from 1 to 15 dB compared with signal depending on the design of the receiver and on the strength of the original noise from atmospheric sources. It is that final combination of noise and signal that is digitized and processed.

In the discussion concerning noise, the bandwidth B of the receiver was mentioned. At one or more stages within the radar system, the received signal is being filtered to select only the frequencies for which signal power is expected. Because the received signal fluctuates in amplitude and phase, it is not a wave with a single frequency f, but is rather made of a distribution of frequencies centered on f but with a bandwidth $1/\tau$ (Fig. 13.3). To properly measure that reflected signal, the receiver must listen to that range of frequencies and therefore have a bandwidth that is at least of the order of $1/\tau$. However, since extending the bandwidth beyond $1/\tau$ has the negative effect of merely increasing the noise being received, receivers are generally designed to have bandwidths of approximately $1/\tau$. For a 1-μs transmit pulse, this would correspond to 1 MHz.

To maximize the radar's ability to detect a signal, the ratio of the signal P_r to the noise P_N must be maximized. Prior to the contribution from the receiver, if the bandwidth of the receiver is $1/\tau$, the signal-to-noise ratio becomes

$$\frac{P_r}{P_N} \approx \frac{1.22^2 0.55^2 10^{-18} \pi^7 c}{1024 k \log_e 2} \frac{P_t \tau^2 D_a^2}{\lambda^4} \frac{T(0,r)^2}{T_A r^2} |K|^2 Z. \tag{13.5}$$

Compared with the radar equation (3.2), this signal-to-noise equation reveals that lengthening the transmit pulse has more benefits than originally considered, because a longer pulse increases the sensitivity on transmit as well as permits the use of a narrower receiver bandwidth, reducing the noise. This explains why, for example, the clear air mode of the

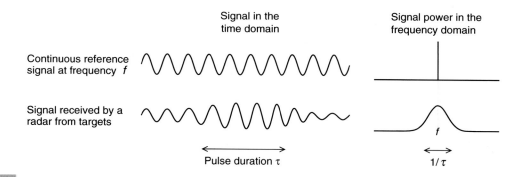

Signal in the
time domain

Signal power in the
frequency domain

Continuous reference
signal at frequency f

Signal received by a
radar from targets

Pulse duration τ $1/\tau$

Figure 13.3.   Illustration of the bandwidth of the radar signal. On the left, two signals are shown, a reference signal at the radar transmit frequency f and the signal received by the radar as a function of time. As the transmit pulse moves from one range to the next, the signal reflected by targets and received by the radar changes amplitude and phase. This results in a signal that appears to fluctuate around the transmit frequency, the rate of fluctuations being dictated by the pulse duration τ, or by how quickly the radar samples a totally different volume. On the right, the resulting power spectra of the two signals are shown. While the reference signal has only one frequency f, the signal received from targets shows a distribution of frequencies centered on f but with a bandwidth close to $1/\tau$.

WSR-88D radars in the United States with its longer transmit pulse has considerably more sensitivity than the precipitation mode, at the cost of some range resolution determined by $c\tau/(2n)$.

13.1.4 From the receiver to the display

Depending on the age of the radar system, the components after the receiver can have many variations (Fig. 13.1). In systems with modern electronics, the analog IF signal is digitized or converted into sets of numbers from which the component I ($=A \cos(\varphi)$) in phase with a reference signal and the component Q ($=A \sin(\varphi)$) in quadrature with that same reference signal are computed, A being the amplitude of the signal and φ the phase difference with respect to the reference (Fig. 13.4). These basic data, known as Level 1 or (I, Q) data, can then be further processed by either general-purpose or specialized computers.

By convention, we consider that the processing of the digital signal is done in two distinct steps. The first step is referred to as the signal processing. It sometimes deals with the removal of artifacts contaminating the radar data such as clutter suppression, and then performs the computation of the basic quantities such as power received and Doppler velocity, also known as Level 2 data. Appendix A.5 introduces some of the signal processing algorithms used to compute reflectivity and Doppler velocity. The second processing step is the product generation, where the Level 2 data are further processed to generate images and products of meteorological use, referred to as Level 3 data or products. As computer systems become more powerful and signal processing is increasingly performed on general-purpose computers, the distinction between signal processing and product generation may become less clear.

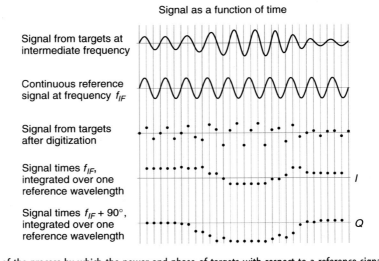

Signal as a function of time

Signal from targets at intermediate frequency

Continuous reference signal at frequency f_{IF}

Signal from targets after digitization

Signal times f_{IF}, integrated over one reference wavelength — I

Signal times $f_{IF} + 90°$, integrated over one reference wavelength — Q

Figure 13.4. Illustration of the process by which the power and phase of targets with respect to a reference signal are determined. The signal from targets (top waveform) whose frequency is centered on $f_{IF} = f - f_{LO}$ is digitized at regular time intervals to obtain a series of numbers (third row). These numbers are then multiplied with a reference signal at frequency f_{IF} (second row), and integrated over one (or several) wavelengths of the reference signal, to obtain $I = A \cos\varphi$ (fourth row). The same is done but with a reference signal 90° out of phase to obtain $Q = A \sin\varphi$ (last row). See Appendix A.5 for the mathematical basis behind this process.

13.2 The weather echo and its fluctuations

If there is only a single target in the direction pointed by the radar, the complex signal measured at a time t after the beginning of the transmit pulse is

$$I(t) + iQ(t) = A\exp[i2\pi f(t - 2nr/c)]W_t(t - 2nr/c), \tag{13.6}$$

where i is the square root of -1 (see Appendix A.5 for a refresher on complex numbers), A is the amplitude of the signal from that target at range r of a radar with a transmit frequency f, n is the refractive index along the path, and $W_t(t-2nr/c)$ is a weighting function in time describing the perceived shape of the transmit pulse, assumed here for simplicity to be 1 when $-\tau/2 \le t - 2nr/c < \tau/2$, and 0 elsewhere. If multiple targets are present, the signal becomes

$$I(t) + iQ(t) = \sum_j A_j\exp[i2\pi f(t - 2nr_j/c)]W_t(t - 2nr_j/c), \tag{13.7}$$

where the signal is summed over all targets identified by the subscript j. The power measured from that signal is proportional to the product of the signal with its complex conjugate, leading to

$$P_r \propto I^2 + Q^2 = \sum_j \sum_k A_j A_k W_j W_k \cos[4\pi fn(r_j - r_k)/c], \tag{13.8}$$

where both j and k are indices to index amplitudes and ranges of individual targets, and $W_t(t-2nr_j/c)$ and $W_t(t-2nr_k/c)$ have been written as W_j and W_k to simplify the notation. Recalling that $\lambda = c/(nf)$, we get

$$P_r \propto \sum_j \sum_k A_j A_k W_j W_k \cos\left[4\pi(r_j - r_k)/\lambda\right]. \tag{13.9}$$

Equation (13.9) reveals that when two targets j and k are in phase, or when $(r_j-r_k)/(2\lambda)$ is an integer, the signal is enhanced, while when targets are out of phase, such as when $(r_j-r_k)/(2\lambda) = 1/2$, the signal is reduced. We can separate (13.9) into two components, one that considers the signal arising from targets independently of one another ($j = k$), and one that describes the effect of the interference between targets ($j \neq k$):

$$P_r \propto \sum_j A_j^2 W_j^2 + \sum_j \sum_{k \neq j} A_j A_k W_j W_k \cos\left[4\pi(r_j - r_k)/\lambda\right]. \tag{13.10}$$

The first term is proportional to the number of targets and to the square of their individual signal amplitude; it is the quantity that can be linked to individual target properties such as their scattering cross section (2.6) from which the radar reflectivity factor (3.1) is derived. The second term, which can be at times much larger than the first, is related to how these targets are distributed in space. If all target distances (r_j-r_k) are equally likely, as would be expected if targets are randomly distributed in space, then the second term tends toward zero when averaged over a sufficiently long time. The resulting probability distribution of received power $p(P_r)$ follows a negative exponential function such that

$$p(P_r) = \frac{\exp(-P_r/\overline{P_r})}{\overline{P_r}}, \tag{13.11}$$

where $\overline{P_r}$ is the average of P_r. From (13.11), it can be derived that a single power measurement has an uncertainty of $\overline{P_r}$, or 100%, and that multiple independent power measurements are required to improve our estimate of average received power, and hence of reflectivity. Equation (13.10) illustrates that the relative distance between targets (r_j-r_k) must change sufficiently between each measurement, of the order of $\lambda/4$ on average, for the second term to take a very different value and hence be considered an independent measurement.

An interesting problem arises if all target distances (r_j-r_k) are not equally likely within the radar sampling volume defined by the pulse weighting function W. In particular, if targets tend to be clustered, smaller (r_j-r_k) would be more likely to occur than larger (r_j-r_k), in which case it cannot be guaranteed that the second term vanishes to zero on average. There is evidence from radar data that, at scales larger than the volume sampled by the radar pulse, precipitation patterns are organized as clusters, cells being embedded in lines or groups of cells, themselves organized in weather systems. When turbulence is sufficient, one can visually observe that raindrops are clustered down to scales of the order of a meter or smaller (see supplement e13.3 or http://www.youtube.com/watch?v=smcfdh_q7SY). It should be noted that it is this process of clustering organized by turbulence, but acting on temperature and moisture fields (Fig. 2.3), that gives rise to the Bragg scattering and the clear air echoes generally observed with longer wavelength radars. At small scales, for precipitation-size targets, two processes oppose each other. On the one hand, as was so poetically written by Richardson in 1922, *"Big whirls have little whirls, which feed on their velocity; and little whirls have lesser whirls, and so on to viscosity"*: turbulence cascades from large scales to small scales, and it is that turbulence that stretches and folds all meteorological fields into patterns that have structures down to millimeter scales. On the other hand, hydrometeors have different fall speeds, as a result of which any cluster of hydrometeors tends to quickly spread vertically, leading to a process similar to vertical mixing that destroys clusters at small scales; it also leads to the spreading of precipitation trails that are commonly observed on vertically pointing radars (Figs. 4.5 and 9.5). Note that this second process is specific to hydrometeors like snow and rain and does not apply to clouds, temperature, and humidity. The magnitude of targets clustering and what effect that clustering has on radar measurements remain uncertain, the literature on this subject often being confusing or misguided. We do believe however that it should not exceed a few dBs in rain under most circumstances; otherwise a mismatch between estimates from radar and from raindrop imager should have been apparent by now.

13.3 Representativeness of measurements

At each range gate, azimuth, and elevation, radar makes measurements of half a dozen fields. Each measurement of atmospheric properties is made within the volume sampled by the radar pulse and the antenna illumination pattern. But this sampling volume does not have a simple shape. In range, it is determined by the pulse length and how it is partly deformed by filters inside the receiver. In azimuth and elevation, it is shaped by G^2, the square of the gain of the antenna, a complicated weighting function that has no sharp boundaries, and which is further smoothed by the antenna movement during the time the measurement is taken. Ideally, meteorologists would like to have measurements that are made at a point, or that represent averages over a volume with simple boundaries such as a sphere, a cube, or a volume with the shape of box. But that is not what radar provides. Hence when we make the assumption that a radar measurement corresponds to what would have been measured at a point or averaged over a simple volume, we make an error of interpretation that is known as the representativeness error. The magnitude of representativeness errors depends on the spatial variability of the field being sampled as well as on the mismatch between the real and assumed sampling volume. These errors can then have a variety of repercussions on how well we can use or further interpret the measurements by radar.

13.3.1 Representativeness errors for reflectivity

Representativeness errors occur when measured fields have large variability at the scale of either the real or the assumed sampling volume. This description fits all fields measured by radar, reflectivity in particular. The rich Fig. 13.5 will be used to gradually illustrate this issue.

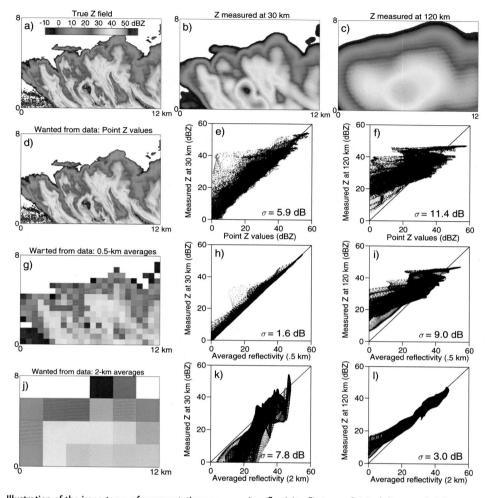

Figure 13.5. Illustration of the importance of representativeness errors in reflectivity. First row: Original distance–height section (a), assumed to be at a constant range, and how it is observed by a radar (b) at 30 km and (c) at 120 km. Second row: Assuming the wanted field to be point values, the wanted field is shown in (d), with (e) and (f) being scatterplots of estimated vs. true point reflectivity made at 30- and 120-km range, respectively. Third row: Similar to the second row, but with the wanted field being 0.5-km averages. Fourth row: Similar to the second row, but with the wanted field being 2-km averages. Think of this plot as if it were a table, with the top row showing the measured fields, the left-most column showing the wanted fields, and the scatterplots at each row–column showing the comparison between each measured and each wanted field. On the scatterplots, the average error (standard deviation) between measured and wanted reflectivity is indicated.

Consider the original high-resolution reflectivity field in Fig. 13.5a collected by a vertically pointing radar. Let us suppose that it is a true field at constant range, and simulate what would be measured by scanning radars at different ranges. Figure 13.5b and 13.5c illustrates what a radar like the WSR-88D (with a 1° beam and scanning over 0.5° azimuth) would observe at 30 and 120 km if it had observations at a very high data resolution in azimuth and elevation. As expected, the storm's structure is smoothed by the radar beam, and as a result much of the small-scale detail disappears at far range. However, systematic changes are even more important. At far ranges, echo patterns are spread vertically, as strong echoes in the bottom of the beam appear as if they were weaker echoes in the center of the beam. This incomplete or partial beam filling can lead to misinterpretations of the storm echo top and structure. Furthermore, in reality, only a subset of these points would be observed, one per elevation and per 0.5° azimuth which further complicates the representativeness issue.

Let us evaluate representativeness errors using the same example. If we assume that the reflectivity measured is representative of the reflectivity at the center of the beam, Fig. 13.5e and 13.5f illustrates the magnitude of the errors that would be made at 30- and 120-km range, respectively. For this event, representativeness errors at 30 km are of the order of 6 dB or a factor 4 in reflectivity *on average*, climbing to 11 dB at 120-km range. In addition, as the mismatch between the sampling volumes increases, weak reflectivity values on the high-resolution image are increasingly overestimated by the radar measurement, while strong reflectivity values become increasingly underestimated. Even when a better representativeness assumption is made, such as trying to match a cube with sides of 0.5 km to the pattern observed at 30 km ($\sin(1°) \times 30$ km ≈ 0.5 km), the errors are smaller but still exceed typical measurement errors. And when the expected and real sampling volumes are mismatched (e.g., Fig. 13.5i and 13.5j), representativeness errors are often huge.

An important conclusion from this exercise is that it will always be difficult to pair observations from one radar with observations from other instruments or estimates from a model that have different viewing or averaging geometry. This is the case when comparing radar and gauge precipitation estimates to derive reflectivity–rain rate relationships, when performing multiple Doppler analyses using two radars, or when constraining model-derived estimates of precipitation with radar observations. For example, it is essential to quantify these representativeness errors when assimilating radar data into models.

13.3.2 Representativeness errors for other radar fields

Reflectivity is not the only field affected by representativeness errors but because the dynamic range and the spatial variability of other fields tend to be smaller, the resulting errors are in general less significant, though not enough to be ignored. There is, however, an added complication: all other radar measurements are reflectivity weighted. For example, if a radar observes two targets in the center of the beam, a small one moving at 2 m/s and a large one moving at 4 m/s, what velocity will the radar measure? The answer is not 3 m/s, but something larger, as the target with the strongest echo will dominate the signal.

This has at least two implications. First, fields such as Doppler velocity, differential reflectivity, and differential phase shift are more representative of the properties of larger targets than of smaller ones. For heavy rain for example, while the diameter D that contributes the most to the rain mass (the peak in $N(D)D^3$) is of the order of 2 mm, the diameter that contributes the most to reflectivity (the peak in $N(D)D^6dD$), and hence to other radar-measured fields, is about twice as large. Second, the properties of targets within portions of the beam that have stronger reflectivity may dominate over the properties of targets at the center of the beam if strong reflectivity gradients are present. For example, the winds that would be measured over the true echo top in Fig. 13.5c actually occur below the center of the beam, as a result of which the Doppler velocity measured around 8 km altitude actually originates from below 7 km. In that case, interpreting the radar measurement as being representative of what happens at 8 km would be a mistake, one that would then affect further data interpretation such as the estimation of the vertical shear as well as data assimilation techniques.

The reflectivity weighting of target properties, especially in the presence of strong gradients, can lead to unexpected measurements such as negative K_{dp} behind storm cells. Figure 13.6 describes how the measured Φ_{dp} may decrease locally even though the Φ_{dp} along individual ray or beam fraction increases with range.. This process occurs without any contribution from a differential backscattering phase delay δ_{co}, which could also easily occur in heavy precipitation, but simply because the contribution from each ray to the total signal changes with the reflectivity encountered. This diversification of Φ_{dp} for each ray will also result in a reduction of ρ_{co} downrange of such strong precipitation gradients, a phenomenon generally attributed to incomplete beam filling.

13.3.3 Minimizing representativeness errors

If measurements from one radar have to be combined with those from other instruments, it becomes essential to minimize the representativeness errors. Figure 13.5 nicely illustrates how the magnitude of these errors diminishes when the size and shape of the volume

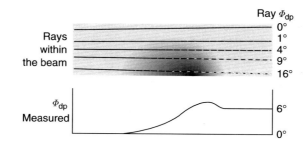

Figure 13.6. Schematic explanation of how negative K_{dp} can be measured in the presence of reflectivity gradients within the beam. On top, a beam represented by five rays of equal gain grazes a reflectivity maximum illustrated by darker gray shades. As a result, each ray is delayed differently, the magnitude of the delay being indicated by the fraction of white dash on each ray, with the total given on the right. But even though the Φ_{dp} of each ray is systematically increasing with range, the Φ_{dp} measured by the radar is locally dominated by the ray with the highest reflectivity, which is also the ray with the most delay. As a result, the plot below of reflectivity-weighted Φ_{dp} measured as a function of range shows decreasing Φ_{dp} behind the reflectivity gradient.

sampled by both instruments become similar. Because it is generally easier to degrade the resolution of measurements than to increase it, the simplest solution is to smooth or combine neighboring measurements of the higher resolution field to mimic the weighting function of the lower resolution field. However, this course of action is not possible when combining measurements from two radars that have complex and incompatible beam shapes, or when assimilating data from radar and another instrument or model grid that has a weighting function with a very different shape. In this case, degrading the resolution of both data sets until their averaging volume becomes sufficiently similar remains the only solution. Such lessons were learned, for example, when trying to make multiple Doppler analyses: smoothing patterns, or not trying to take into account small-scale patterns, generally led to better results because smoothing minimized mismatches caused by representativeness errors.

13.4 Where do we go from here?

What this broad overview illustrates is that proper interpretation and use of radar data requires that we pay attention to the details of how measurements are being made and interpreted. Radar can be a very difficult instrument in that respect. For qualitative and semiquantitative applications of interest to most users of radar data, such issues are largely irrelevant, which is why I chose to mention them only at the end of this book. But the moment radar is to be used quantitatively, a lot of attention must be put into these considerations, much more than has been the norm until now. All that being said, radar remains the best instrument at our disposal to make measurements about storms, especially at subsynoptic scales, and we therefore have no choice but to confront these challenges.

But that should not be a reason to despair or give up. Since its beginnings in the 1940s, the use of radar for meteorology has undergone tremendous progress, and will continue to do so in the future. We must simply accept to face the challenges, as ignoring them is the surest way to ultimate failure. Research papers, preprints of the many radar meteorology and hydrology conferences, and the occasional review books testify of the dynamism of our field. These also constitute vast treasure troves that deserve to be explored.

Despite possible future progress, it is hard to envisage the physical basis of any remote sensing technology that could replace radars and provide similar information about storms. The use of radar for meteorology is therefore here to stay. What shape and form it will take in the future is however left to anyone's imagination.

Appendix A   Mathematics and statistics of radar meteorology

Many branches of meteorology rely heavily on mathematics to help explain and describe phenomena. Most, especially those rooted in atmospheric dynamics, primarily use calculus and differential equations; as a result, these mathematical concepts are well taught and often used in meteorology programs and courses. In many ways, the mathematical foundations of instrumentation in general and of radar in particular are more basic, being rooted in geometry, algebra of real and complex numbers, and introductory statistics. But since these do not get applied as much in meteorology programs, they tend to be forgotten. Furthermore, geophysical fields such as winds and rainfall pose special challenges to their analysis that are often ignored in introductory statistics courses, leading to mistakes that are too often observed even in scientific publications. This section introduces some of these topics by focusing on particular radar research problems that require a specific mathematical treatment or statistical approach to study them, and use these problems as means to introduce the relevant mathematical or statistical concepts. It will, however, not be a complete surrogate to proper reference texts in mathematics and statistics, but should hopefully help the reader ask the right questions and find the relevant material in these books. This appendix also gives me the opportunity to complement or introduce topics that could not be easily covered in previous chapters.

A.1 Geometry of ground-based radar measurements

To interpret or simulate radar measurements, it is essential to understand the complexity of the geometry of radar measurements. This complexity is due, among others, to the fact that both the radar measurements and the Earth-centric coordinate system follow different near-spherical geometry. On the one end, radar makes measurements in range–azimuth–elevation (r, ϕ, θ) along beams that have widths and that bend due to propagation in the atmosphere. On the other end, the position and axis of measurement must be reported with respect to an Earth-centric coordinate system such as latitude–longitude–height (L_p, l_p, z), where the pointing direction of axes such as "east" and "up" changes from location to location because of the curvature of the Earth. As a result, a beam starting from the radar toward the east at a certain elevation angle will hit targets with an axis that is not exactly from west to east and with a generally higher elevation angle than pointed by the radar. Those slight changes are particularly important to the proper simulation of radar observations given a model field.

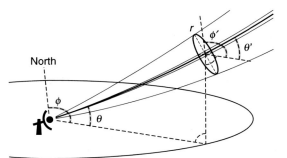

Figure A.1. Illustration of the path of a radar beam when the radar points at azimuth ϕ, and elevation θ. As a result of Earth curvature and beam bending, at range r, the angle between the trajectory of the beam and the local horizontal plane is θ' while the angle with respect to the local North is ϕ'.

A.1.1 Radar measurements on a single ray

Let us first consider the radar beam to be infinitely narrow, keeping beam width considerations for later. Figure A.1 shows the geometry of that measurement and illustrates the meaning of several of the symbols used. Focusing on the horizontal first, for a given radar at latitude L_{rad}, longitude l_{rad}, and altitude z_{rad}, the latitude L_p and longitude l_p of the pulse as a function of azimuth ϕ, elevation θ, and range r can be determined. Using spherical geometry applicable for ranges r much smaller than the radius of the Earth a_e, we obtain (e.g., Williams 2014)

$$L_p \approx \sin^{-1}\left[\sin\left(L_{rad}\right) \cos\left(\frac{r\cos\theta}{a_e + z_{rad}}\right) + \cos\left(L_{rad}\right) \sin\left(\frac{r\cos\theta}{a_e + z_{rad}}\right) \cos\phi \right] \qquad (A.1a)$$

and

$$l_p \approx l_{rad} + \tan^{-1}\left[\frac{\cos\left(L_{rad}\right) \sin\left(\frac{r\cos\theta}{a_e+z_{rad}}\right) \sin\phi}{\cos\left(\frac{r\cos\theta}{a_e+z_{rad}}\right) - \sin\left(L_{rad}\right) \sin\left(L_p\right)} \right]. \qquad (A.1b)$$

As a result, the azimuth angle of propagation ϕ' with respect to the local North direction at range r becomes (Fig. A.1)

$$\phi' \approx \pi + \tan^{-1}\left[\frac{\sin\left(l_{rad} - l_p\right) \cos\left(L_{rad}\right)}{\cos\left(L_p\right) \sin\left(L_{rad}\right) - \sin\left(L_p\right) \cos\left(L_{rad}\right) \cos\left(l_{rad} - l_p\right)} \right]. \qquad (A.2)$$

At 45°N latitude and 120-km range, for an azimuth of 90° and at low elevation, ϕ' becomes approximately 91° and the latitude will have changed by the equivalent of a couple of kilometers (see supplement eA.1 for a simple spreadsheet implementing these equations). A wind from the south at that location will therefore have a nonzero radial velocity.

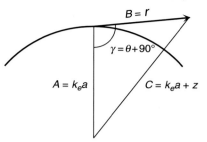

Law of cosines: $C^2 = A^2 + B^2 - 2AB \cos \gamma$

Figure A.2. Illustration of the derivation of the altitude z pointed by the radar at range r and elevation θ using the law of cosines applied to the geometry of the effective earth assuming a radar altitude z_{rad} of 0.

Conversely, to determine the azimuth and range from the radar to reach the final latitude L_p and longitude l_p, the following may be used:

$$\phi \approx \tan^{-1} \left[\frac{\sin \left(l_p - l_{rad} \right) \cos \left(L_p \right)}{\cos \left(L_{rad} \right) \sin \left(L_p \right) - \sin \left(L_{rad} \right) \cos \left(L_p \right) \cos \left(l_p - l_{rad} \right)} \right] \tag{A.3a}$$

and

$$r = a_e \cos^{-1} \left[\sin \left(L_{rad} \right) \sin \left(L_p \right) + \cos \left(L_{rad} \right) \cos \left(L_p \right) \cos \left(l_{rad} - l_p \right) \right]$$

$$= 2a_e \sin^{-1} \left[\sqrt{ \sin^2 \left(\frac{L_{rad} - L_p}{2} \right) + \cos \left(L_{rad} \right) \cos \left(L_p \right) \sin^2 \left(\frac{l_{rad} - l_p}{2} \right) } \right] \, , \tag{A.3b}$$

the second equation of (A.3b) being mathematically equivalent to the first but less subject to numerical truncation errors for short ranges.

The elevation also changes with range, both due to the curvature of the Earth and the bending of the propagating beam caused by the vertical gradient of the refractive index of air dn/dz. This affects both the altitude z observed by the radar and the elevation angle with respect to which Doppler velocity and polarimetric quantities sensitive to target shape are measured. A traditional approach to estimate the height z and the elevation angle of propagation θ' is to use the law of cosines (Fig. A.2) combined with a $k_e = 4/3$ Earth approximation (2.13), leading to (e.g., Ge *et al.* 2010)

$$z = \left\{ r^2 + \left[\left(k_e a_e + z_{rad} \right)^2 + 2r \left(k_e a_e + z_{rad} \right) \sin \theta \right] \right\}^{\frac{1}{2}} - k_e a_e \tag{A.4a}$$

and

$$\theta' = \theta + \tan^{-1} \left(\frac{r \cos \theta}{k_e a_e + z_{rad} + r \sin \theta} \right), \tag{A.4b}$$

with

$$k_e = \frac{1}{1 + \frac{dn}{dz}(a_e + z_{rad})} \simeq \frac{4}{3} \quad \text{for } \frac{dn}{dz} \approx -40 \times 10^{-9} \text{m}^{-1}. \tag{A.4c}$$

Equations (A.4a) and (A.4b) are somewhat simplistic because they are based on an assumption of constant dn/dz, often assumed to be the one near the surface. To accurately compute the altitude pointed by the radar, (A.4) must be solved iteratively for small-range segments. The spreadsheet in supplement eA.1 performs a limited version of such a calculation using an exponential dn/dz profile. More sophisticated approaches that better take situations like trapping or negative elevation angles into account have also been developed (Zeng *et al.* 2014).

A.1.2 Radar scans and the structure of radar data

The geometry of measurements, combined with the scanning strategy of radars, also dictates the structure used to archive radar data. Even if it seems out of place, a brief description of this structure is useful here to introduce concepts needed later.

Operational radars generally perform a regular scanning routine, such as making the same set of 18 PPIs at different elevations every 5 min. This regular pattern is called a volume scan, while the individual PPIs or RHIs that compose this volume scan are generally referred to as sweeps. Over the time taken by the radar to complete a sweep, multiple transmit pulses are fired. The signal is digitized at regular time intervals after each transmit pulse to obtain information in range; each location in range where data are digitized is known as a range gate. Raw range gates are typically spaced by $c\tau/(2n)$, or of the order of every 150–250 m for scanning radars; sometimes though, measurements are averaged in range prior to being archived, as a result of which data cells are available every 500 m to 1 km. The ray of information ensuing from each transmit pulse is referred to as a radial that is at a specific azimuth and elevation angle. Generally, many radials are combined over 0.5° or 1° of azimuth for a PPI, or over 0.2° to 0.5° of elevation for an RHI, prior to archiving. Therefore, the data for a typical PPI will be made up of 360–720 radials, themselves made up of a few hundreds to a few thousands of data cells at different ranges.

Figure A.3 illustrates the typical organization of an archive of scanning radar data. Generally, each volume scan is archived in a single file, as that file conveniently contains the information needed to generate all the radar products at that particular time except for accumulations and echo tracking products that necessarily need data from multiple volume scans. Each volume scan structure generally starts with a block of information, known as a volume scan header, that provides contextual data such as radar characteristics and location, time, radar health information, etc., as well as information on how many sweeps to expect. Then follows a number of sweep structures that contain the data for each elevation. Depending on the system, each sweep structure describes the information on one measured field such as reflectivity, or on all measured fields. Each sweep structure is itself made of a sweep header and of radial structures, themselves made of a radial header and data cell structures. Because, on average, radar echoes are observed in less

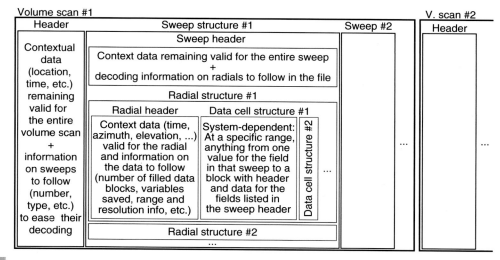

Generic structure of archived scanning radar data.

than 5% of the cells, data are generally not archived for all range gates; often, only data cells with echoes have been archived, the others being assumed to have no echo. Within each data cell block or structure, values for each field are often encoded as integer values V that must be decoded using Measured_Field = Slope $\times$ V + Offset, the numbers for Slope and Offset for that particular field being listed in one of the headers. Generally, power-based fields are archived in units of dB or dBZ, velocity-based fields in meters per second, and phase-based fields in degrees.

A.1.3 Measurement simulations

To compare radar data with information from other sources, it is often necessary to simulate what a radar would have observed under given circumstances. The first step in such simulations is to reproduce the geometry of the measurement using some or all of the equations above. These are sufficient to allow the simulation of measurements made by infinitely narrow beams and short pulses. For proper measurement simulations, sampling volume considerations and the smoothing done in range and azimuth prior to archiving must be added.

Radar measures the average properties of targets over a volume that is determined by two independent factors: in range, it is shaped by the transmit pulse and receiver characteristics; in azimuth–elevation, it arises from the beam pattern and the smearing caused by the antenna movement during the tens of milliseconds over which measurements from successive transmit pulse are averaged. Since the volume does not have sharp boundaries, we define two independent weighting functions that describe the relative sensitivity of the radar to the reflectivity of each volume element. In range, the weighting functions $W_r(\delta r)$ for each cell centered at range r and with a range deviation δr with respect to that center are strictly speaking a function of the length and shape of the transmit pulse, receiver filter and digitizer characteristics, and how many raw measurements were combined in range to obtain the radar measurement that needs to be simulated. In general, when multiple measurements are combined in range, $W_r(\delta r)$ can often be simplified to a step function and assumed to be $1/\Delta r$ for δr within $[-\Delta r/2, \Delta r/2]$, and 0 elsewhere. If the details of how different hardware components affect the perceived pulse shape become important, then $W_r(\delta r)$ becomes more complicated to specify (Torres and Curtis 2011). Under most circumstances though, because the smoothing in range is smaller than that in azimuth or elevation, it is much more important to correctly calculate the smoothing in angle than the one in range, as a result of which $W_r(\delta r)$ can be simplified to a step function without much loss of accuracy.

The angular weighting function with respect to the beam axis $W_a(\delta\phi, \delta\theta)$ is the sum of the square of the gain G of the antenna (13.2) over each transmit pulse:

$$W_a(\delta\phi, \delta\theta) = \frac{1}{n_p} \sum_{j=1}^{n_p} G^2(\phi_j - \phi, \theta_j - \theta), \tag{A.5}$$

where n_p is the number of transmit pulses whose signal was combined in the data archived (generally anywhere from 16 to 128) and ϕ_j and θ_j are the azimuth and elevation pointed by the antenna when each transmit pulse was fired. In the absence of detailed knowledge on ϕ_j and θ_j, for PPIs, one can assume $W_a(\delta\phi, \delta\theta)$ to be

$$W_a(\delta\phi, \delta\theta) = \frac{1}{\Delta\phi} \int_{\phi_d=\overline{\phi}-\Delta\phi/2}^{\overline{\phi}+\Delta\phi/2} G^2(\phi_d - \phi, \overline{\theta} - \theta)d\phi_d, \tag{A.6}$$

with $\overline{\phi}$ and $\overline{\theta}$ being, respectively, the average or center azimuth and elevation pointed by the antenna, and $\Delta\phi$ is the azimuth interval over which echoes are averaged. In general, to avoid the largest errors, it is more important to consider the beam pattern in elevation than in azimuth as the strongest reflectivity gradients and the largest biases in measurement simulations occur when the beam pattern in elevation is ignored.

If we account for the beam pattern and for attenuation, given a true field of reflectivity Z, the equivalent reflectivity factor Z_e observed by the radar at elevation θ, azimuth ϕ, and range r is

$$Z_e(\theta, \phi, r) = \frac{1}{|K_w|^2} \int_{\theta_d} \int_{\phi_d} W_a(\theta_d - \theta, \phi_d - \phi) \int_{r_d} W_r(r_d - r)T(0, r_d)^2 |K|^2 Z(\theta_d, \phi_d, r_d)dr_d d\phi_d d\theta_d, \tag{A.7}$$

where $T(0, r_d)$ is the transmittance of the atmosphere along the path between the radar at range 0 and the volume element considered at range r_d, $|K|^2$ is the average dielectric constant of the scatterers, while $|K_w|^2$ is the dielectric constant of liquid water. For measurement simulations, for each (θ_d, ϕ_d, r_d) combination, (A.1) and (A.4) are used to determine the path taken by the radar waves and the resulting latitude, longitude, and height at destination, while computing the transmittance of the atmosphere along that path. At the end of the path, the value of Z is determined from the model or other fields whose measurements we are trying to simulate, as well as the dielectric constant K^2 of the targets considered. The weighting functions W_a and W_r must then be considered to obtain the resulting contribution to the final equivalent reflectivity factor from that particular (θ_d, ϕ_d, r_d) combination. This must then be repeated for all (θ_d, ϕ_d, r_d) combinations whose $W_a W_r$ is large enough to provide a meaningful contribution to the final result.

To simulate other radar-measured fields, the same process can be used, remembering that the contribution from each ray is also a function of the strength of the returning signal. A further complication arises for fields whose magnitude depends on elevation such as Doppler velocity v_{DOP} and measured Z_{dr}. The elevation to be considered for their computation is not the elevation angle θ of the antenna but the angle of the beam θ' with respect to the horizon defined in (A.4b). Hence, to compute the Doppler velocity v_{DOP} ignoring effects introduced by signal processing issues, we should use

$$v_{DOP}(\theta, \phi, r) = \frac{\int_{\theta_d}\int_{\phi_d} W_a(\theta_d - \theta, \phi_d - \phi) \int_{r_d} W_r(r_d - r)T(0, r_d)^2 |K|^2 Z(\theta_d, \phi_d, r_d)\mathbf{s} \cdot (\mathbf{v} - w_f\mathbf{k})dr_d d\phi_d d\theta_d}{|K_w|^2 Z_e(\theta, \phi, r)}$$

(A.8)

where $(\mathbf{v} - w_f\mathbf{k})$ is the vectorial velocity of targets pushed by a 3-D wind $\mathbf{v}$ and falling with a fall speed w_f, $\mathbf{s}$ is a unit vector perpendicular to the radar phase fronts such that $\mathbf{s} = \cos\theta' \sin\phi'\mathbf{i} + \cos\theta' \cos\phi'\mathbf{j} + \sin\theta'\mathbf{k}$, $\mathbf{i}, \mathbf{j}$, and $\mathbf{k}$ being unit vectors pointing, respectively, east, north, and up, while θ' and ϕ' are defined in (A.4b) and (A.2). Failing to take into account both the beam pattern and the reflectivity field leads to biased wind simulations as radar-observed winds tend to come from a lower altitude than what is expected by the position of the center of the beam.

In order to simulate what quantities the radar would observe using model fields, it is generally necessary to infer reflectivity from mass and maybe number concentration as models do not commonly compute reflectivity. Some approximations linking reflectivity to other quantities of meteorological interest are provided in supplement eA.2. Note that whatever the microphysics used, these relationships are relatively simple approximations that introduce significant errors, errors that also have long correlation distances as microphysical and dynamical processes shaping precipitation evolution are also correlated in space. These errors and their spatial correlation must somehow be taken into account in the observational error covariance matrix.

A.2 Operations on reflectivity data

For historical and practical reasons, reflectivity data are generally archived in dBZ or logarithmic units, not in mm^6/m^3 or linear units; so are Z_{dr} and LDR. Once data are read back, the temptation is hence great to manipulate and use reflectivity data in logarithmic units. There are many circumstances for which this is not a good idea, and a few where it is.

To understand why this is the case, it must first be noted that operations done in logarithmic units give different answers than in linear units. If we consider averaging M_s reflectivity measurements for example,

$$\overline{Z} = \frac{1}{M_s} \sum_{j=1}^{M_s} Z_j \geq \left(\prod_{j=1}^{M_s} Z_j \right)^{1/M_s} = 10^{\overline{[\log_{10}(Z)]}}. \tag{A.9}$$

Averaging in linear units computes an arithmetic average, while averaging in logarithmic or dB units computes a geometric average, the two yielding different results unless all samples have identical value. For example, the average of a 0 dBZ echo (1 mm^6/m^3) and a 20 dBZ echo (100 mm^6/m^3) leads to a 10 dBZ (10 mm^6/m^3) geometric average and a 17 dBZ (50.5 mm^6/m^3) arithmetic average. Geometric averages are smaller than the usual arithmetic averages, the difference between the two becoming larger as the distribution of values to be averaged becomes wider. In parallel, often greater differences in results between linear and logarithmic operations are obtained when two numbers are subtracted as would happen when computing deviations from an average. Therefore, any quantitative use of the result of mathematical operations such as averaging and subtracting must consider the effect of such differences in results. And while averaging and subtracting reflectivity in linear units may have some physical meaning, doing so in logarithmic units generally does not.

Furthermore, reflectivity measurements from hydrometeors can range from zero to millions of mm^6/m^3, zero being the most common value. A challenge faced when using logarithmic units is how to represent and manipulate log(0). Clearly, $-\infty$ is not a useful choice, as any averaging involving $-\infty$ leads to a result of $-\infty$, severely limiting the information that can be obtained from data processing. A way around this problem is to set a minimum value for reflectivity, often chosen to be 0 dBZ. But because a large fraction of the reflectivity field has zero value, the final result of most data processing becomes very sensitive to the choice of that minimum value, and therefore very sensitive to an arbitrary parameter with no scientific merit.

One of the rare circumstances for which the use of dBZ units is advisable is for correlation computations, such as those used to track reflectivity patterns from one radar image to the next. Correlation calculations work best and have quantitative meaning when the quantities to be correlated follow a normal distribution. In that respect, processing reflectivity data in dBZ units is better than in linear units (Fig. A.4): distributions of reflectivity tend to look exponential, and the computed correlation will be primarily determined by the few most intense reflectivity values. In logarithmic units, reflectivity often follows a normal distribution to which one must add a very strong secondary peak for zero values. If zero values can be excluded from such correlations, then correlation computations using dBZ values should provide the best results. The same is true for other variables with similar intensity distributions such as rainfall rates and amounts.

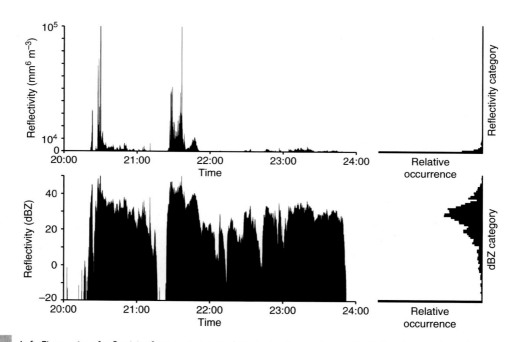

Figure A.4. Left: Time series of reflectivity from a rain event plotted using (top) a linear reflectivity scale and (bottom) a logarithmic reflectivity scale. On the right, the associated histograms of occurrence of different reflectivity are shown. Histograms of reflectivity generally resemble exponentials while those of logarithm of reflectivity look more like normal distributions to which are added a second peak for reflectivity values of zero.

A.3 Regressions and mapping functions

A.3.1 Context

We often have a need to infer an otherwise unmeasured quantity from a measured one. A classic example in radar meteorology is when we try to estimate rainfall from reflectivity measurements. In order to derive relationships between these two quantities, we often have to compute regressions between them.

Because the need for regressions is so prevalent in research, many types of regressions have been developed. They vary according to the functional relationship they use (e.g., linear, polynomial, and power law) and the quantities they are trying to minimize to compute the regression (least squares vs. absolute values, one or two dependent variables, with or without error weighting, etc.). Each type of regression optimizes something different somewhat differently, and these choices have implications on the final result obtained. Hence, most statistics books devote several chapters to that topic, and many online resources give an overview of the breadth of possibilities available. Therefore, I will not try to explain all these variations. Instead, I will illustrate the implications of some of the choices made by using one set of approaches over another. Because the traditional regression problem in radar meteorology involves deriving a power-law relationship between variables that span several orders of magnitude, this is the context used in the discussion to follow.

A.3.2 Insights from a power-law regression experiment

Consider the following simulation: let us start with a map of 1-h accumulation of precipitation (Fig. A.5, left), and assume that this map is the truth we would like to obtain. Then let us displace that map by 2 km to the east, and assume that this is the measurement that we have. As a result of the mismatch in position between "truth" and

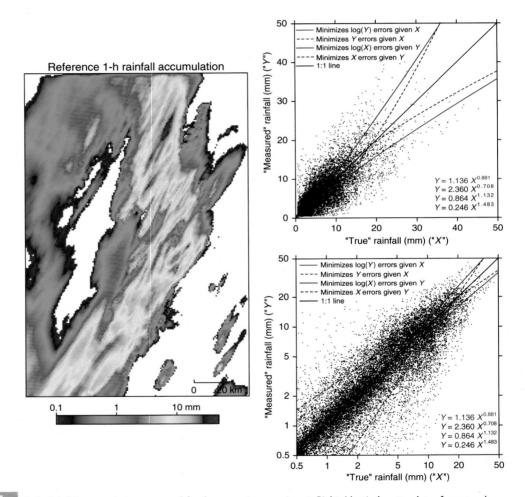

Figure A.5. Left: Rainfall accumulation map used for the regression experiment. Right: Identical scatterplots of measured compared with true rainfall but plotted on two sets of axes, linear axes on top and logarithmic axes at the bottom. Superposed are the results of four power-law regressions made using different assumptions on the quantity to be minimized (errors in Y in reds and errors in X in blues) and the functional dependence of errors (constant magnitude errors in dashed lines and constant fractional error in solid lines).

"measurement," each measurement has an error. In fact, the resulting magnitude of the error in this particular example is very similar to that observed when comparing radar- and gauge-derived accumulations. Each pair of measurement–truth pixels where both truth and measurement exceed 0.5 mm is plotted on the right of Fig. A.5 and will be used to compute regressions. Two scatterplots are shown, one using linear axes and one using logarithmic axes, as each provides a different perspective. Finally, four power-law regressions are computed on these two data sets in an attempt to derive one variable from the other, and these are plotted on top of both scatterplots. These regressions are all made on the assumption that we have one dependent variable Y and one independent variable X, the default model used in most regression algorithms; what is varied is which variable is assumed to be the independent one, and whether the error to minimize for each measurement–truth pair is assumed to be proportional to the rainfall amount or a constant.

The four regression functions f all tried to minimize different quantities, and it is important to understand these differences. Mathematically, within the constraints of a power-law regression, the function plotted as the red solid curve minimized the variance (or difference squared) between $\log(Y)$ and $\log(f(X))$; the darker dashed red curve did the same minimizing the variance between Y and $f(X)$; the solid blue curve minimized the variance between $\log(X)$ and $\log(f^{-1}(Y))$; and finally, the darker dashed blue curve did the same minimizing the variance between X and $f^{-1}(Y)$. To help visualize what this represents, the above corresponds graphically to the following: the red solid curve minimized the distance between itself and points in the vertical direction in the bottom figure, the darker red dashed curve doing the same on the top figure; the blue solid curve minimized the distance between itself and points in the horizontal direction in the bottom figure, the darker blue dashed curve doing the same on the top figure.

A key result of this exercise is the difference in the regressions obtained, the exponent in the power law varying by a factor 2. This simply arose from changing which quantity is optimized. If the goal is to predict true rainfall from measured rainfall, then one of the blue curves should be used, depending on whether we want a relationship that achieves the best relative accuracy for different rain accumulation regimes (dashed line) or one that achieves the best unbiased rainfall overall (solid line). Note that (1) the spread of values for b in regressions in this simple exercise are comparable to that obtained for Z–R relationships in the literature and (2) the temptation is great to use R as the independent variable because relationships are often written as $Z = f(R)$ even though R is the quantity we are trying to predict. Given that exact details on what regression was used are rarely included in short scientific publications, we will likely never know how much of the observed variability in published Z–R relationships is simply due to different error minimization approaches.

What about the one-to-one line $Y = X$? Though it is the relationship we might have expected to find given that X and Y come from the same dataset, it does not minimize the total error variance of going from one data set to the other, which is what traditional regressions are designed to obtain. In other words, it does not provide the smallest errors overall when trying to predict Y given X. However, $Y = X$ is the only function that preserves the dynamic range of values and the probability of occurrence of different rain rates, such as the probability of catastrophic events. To obtain relationships that preserve the probability of rare events, a different type of regression must be used, such as a total least squares regression.

A.3.3 Some dangers of optimized relationships

This discussion raises a complicated issue that is often ignored: the need to properly define what exactly we want to optimize for when deriving relationships between two variables. Regressions are all about obtaining the best result given a set of constraints, such as minimizing the average standard deviation of the error given a power-law relationship. In general, this optimization will be made at the expense of every other property that is not specifically mentioned.

We often have contradictory sets of requirements and make the mistake of only specifying a subset of them. For most applications, the best relationship between what is measured and what we want is the one that minimizes errors. But we also often want to preserve the width of the distribution, and the likelihood of rare events, as those are the important events to correctly measure or predict. Furthermore, the optimum relationship between X and Y may not be the one that will lead to the best derived quantity $f(Y)$ given X. If what we really want is $f(Y)$, then what should be minimized in the regression are the errors in $f(Y)$, not in Y.

For example, the Z–R relationship optimized to obtain the best hourly accumulation is not the one that will provide the best instantaneous rain rates, nor the best 6-h accumulation. In general, as the relative variability around the expected regression line increases, smaller errors are generally obtained by regressions with smaller slopes: such regressions diminish the range of predicted values and avoid outliers that are often heavily penalized for. But rain rates with a diminished range will not yield good results when we accumulate them in an attempt to obtain reliable longer term accumulations, especially if these accumulations are unusually high or low. To obtain the best accumulations, a Z–R relationship that better preserves extremes would perform better.

A similar exercise could have been performed using other relationships, such as deriving reflectivity at the surface from reflectivity measurements aloft using vertical profiles of reflectivity: if multiple profiles are generated for different observed echo intensity aloft, the derived reflectivity at the surface will vary depending on whether profiles were derived to minimize the error on the instantaneous reflectivity estimate, to minimize the 1–h accumulation rainfall error, or to preserve the dynamic range of precipitation aloft or at the surface.

The key conclusion of this discussion is that there is no single relationship between two variables that will always perform optimally all the time. It is therefore crucial to reflect on exactly what we want to optimize for when deriving such relationships, as the result obtained may not be the one we really want.

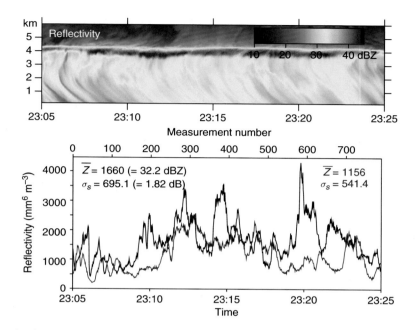

Top: Time–height section of reflectivity measured by a vertically pointing radar during a rainfall event. Bottom: Time series of reflectivity at 1.7-km (black) and 0.3-km (red) altitudes plotted in linear units, 790 series of measurements having been made over this 20-min period.

A.4 Variances, covariances, and autocorrelations

A.4.1 Correlation between two data sets

Consider the two time series of measurements in Fig. A.6. With their succession of maxima and minima with no clear periodicity or structure, they have statistical characteristics that resemble that of many atmospheric and geophysical fields. In this example, they were derived from vertically pointing radar data that sampled a subset of a storm event, but plots such as these could have been made from a variety of atmospheric sensors and look very similar. For both samples, we can compute their sample average $\overline{Z}$ and their sample variance σ_s^2 using

$$\sigma_s^2 = \frac{1}{M_s - 1} \sum_{j=1}^{M_s} (Z_j - \overline{Z})^2 = \frac{1}{M_s - 1} \left[\left(\sum_{j=1}^{M} Z_j^2 \right) - \frac{1}{M} \left(\sum_{j=1}^{M} Z_j \right)^2 \right], \qquad (A.10)$$

where M_s is the number of measurements in the sample. From (A.10), the sample standard deviation σ_s can be computed. For the sample plotted in black in Fig. A.6, $\overline{Z}$ is close to 1660 while σ_s is 695.1. If we chose to express both in dB, the average is simply converted the usual way ($10 \log_{10}(Z)$), while the standard deviation must be converted using (A.11) that computes the uncertainty or standard deviations in one unit given the uncertainty or standard deviation in another unit:

$$\sigma_s[f(X)] = \left[\frac{\partial f}{\partial X}\right]_{\overline{X}} \sigma_s(X), \tag{A.11}$$

where the partial derivative of the conversion function f with respect to the original variable X, here reflectivity, is evaluated at the average value of X. In our case, this leads to

$$\sigma_s[f(Z) = 10 \log_{10}(Z)] = \frac{10}{\overline{Z} \log_e(10)} \sigma_s(Z). \tag{A.12}$$

Via this process, the converted sample standard deviation can describe how variable the reflectivity values are in dB units.

The two time series are of reflectivity at different altitudes but for the same event. We therefore expect them to be correlated: that is, we expect maxima in reflectivity in one sample to often correspond to maxima in the other sample, and similarly for minima. To quantify that correlation, we must first estimate the covariance between these two sets of samples X and Y:

$$\text{cov}_{X,Y} = \frac{1}{M_s - 1} \sum_{j=1}^{M_s} (X_i - \overline{X})(Y_i - \overline{Y}). \tag{A.13}$$

Note that if the samples X and Y are identical, $\text{cov}_{X,Y}$ is $\text{cov}_{X,X}$ and reduces to $\sigma_s(X)^2$, the variance of X. Also be aware that (A.13) tests for covariance assuming a linear relationship between the two samples; if the two samples covary but based on a more complicated relationship, this covariance calculation will not yield a meaningful result. The norm of the sample covariance $|\text{cov}_{X,Y}|$ is bound by $\sigma_s(X)\sigma_s(Y)$, the product of the standard deviation of samples X and Y. The ratio of the covariance and of $\sigma_s(X)\sigma_s(Y)$ is the sample linear correlation coefficient $\rho_{X,Y}$. Hence,

$$\rho_{X,Y} = \frac{\text{cov}_{X,Y}}{\sigma_s(X)\sigma_s(Y)} = \frac{\sum_{j=1}^{M_s}(X_i - \overline{X})(Y_i - \overline{Y})}{\sqrt{\sum_{j=1}^{M_s}(X_i - \overline{X})^2 \sum_{j=1}^{M_s}(Y_i - \overline{Y})^2}}. \tag{A.14}$$

Highly positive or negative correlations between two samples suggest that, given measurements of one sample, the second one could be predicted with some skill. This is particularly important for some applications such as data assimilation where we must rely on measurements or constraints on some fields such as rainfall and horizontal winds to obtain information on unobserved fields such as vertical winds, temperature, and humidity. The possibility of predicting one set of samples given another also forms the basis for regressions and mapping functions.

A.4.2 Lagged correlation

On the radar image of Fig. A.6, we can see that precipitation trails are slanted. Therefore, the best correlation between the reflectivity observations at 1.7 and 0.3 km would not be observed when the two time series are superposed, but rather when one is displaced, or lagged, compared with the other. One can then compute lagged sample covariances where one time series is displaced by l points:

$$\mathrm{cov}_{X,Y}[l] = \left\{ \begin{array}{ll} \dfrac{1}{M_{\mathrm{s}} - l - 1} \displaystyle\sum_{j=1}^{M_{\mathrm{s}}-l} (X_i - \overline{X})(Y_{i+l} - \overline{Y}) & \text{for } l \geq 0 \\[3ex] \dfrac{1}{M_{\mathrm{s}} + l - 1} \displaystyle\sum_{j=1-l}^{M_{\mathrm{s}}} (X_{i-l} - \overline{X})(Y_i - \overline{Y}) & \text{for } l < 0 \end{array} \right\}, \tag{A.15}$$

and from these lagged sample covariances, lagged correlations can be computed using $\rho_{X,Y}[l] = \mathrm{cov}_{X,Y}[l] / [\sigma_{\mathrm{s}}(X)\sigma_{\mathrm{s}}(Y)]$. Note that as the norm of the lag $|l|$ approaches the number of measurements M_{s}, the estimates of covariance and of correlation become unreliable. Lagged correlations were computed from the two time series shown in Fig. A.6 using reflectivity in both linear and logarithmic units and have been plotted in Fig. A.7. Peak lagged correlations are obtained for a lag of 42 points (for linear Z) and 44 points (for logarithmic Z), which correspond in this case to a time lag between the two time series of slightly more than 1 min.

What has just been done in 1-D using vertically pointing radar data can be done in 2-D or in 3-D using scanning radar data. If two images 5 min apart are correlated, the lag or displacement where the peak correlation occurs corresponds to the displacement of the weather echoes over these 5 min; these can then be used to extrapolate storm movement and make very short-term predictions of future storm locations (Chapter 10).

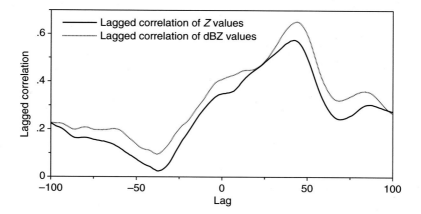

Figure A.7.   Lagged correlations between the two time series shown in Fig. A.6.

A.4.3 Lagged autocorrelation

While the lagged covariance in (A.15) is often computed from two different samples, nothing prevents us from using the same sample to compute lagged autocovariances:

$$
\mathrm{cov}_{X,X}[l] = \left\{
\begin{array}{ll}
\dfrac{1}{M_s - l - 1} \displaystyle\sum_{j=1}^{M_s - l} (X_i - \overline{X})(X_{i+l} - \overline{X}) & \text{for } l \geq 0 \\[3ex]
\dfrac{1}{M_s + l - 1} \displaystyle\sum_{j=1-l}^{M_s} (X_{i-l} - \overline{X})(X_i - \overline{X}) & \text{for } l < 0
\end{array}
\right\}.
\tag{A.16}
$$

As was done using covariances, autocorrelations can be computed from autocovariances. At lag zero, the autocovariance becomes the variance $\sigma_s^2(X)$ of the sample, while the autocorrelation $\rho_{X,X}[0] = \mathrm{cov}_{X,X}[0]/\sigma_s^2(X)$ reduces to 1. But the study of how autocorrelations vary as a function of lag is particularly interesting because of what it teaches us about the sample.

Figure A.8 shows the autocorrelation as a function of the lag l for the time series of reflectivity at 1.7 km. There are three regions of particular interest in this plot: the autocorrelation values within two points of lag 0, the region of near-exponential decrease between lag 2 and 20, and the rest of the curve. Focusing first on the middle region, the rate of decrease of autocorrelation can be used to determine the correlation distance or time of patterns in the field of study. For the reflectivity time series used here, the correlation distance, which corresponds to the point where correlation decreases to $1/e$ or about 0.37, is approximately 25 points or 40 s in this case. At higher lags, the curve can often take complicated shapes,

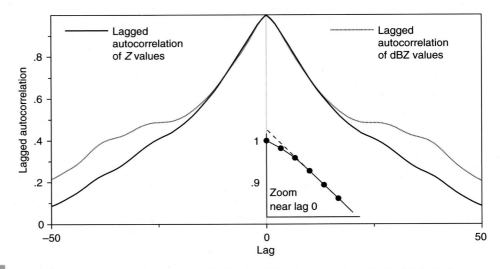

Figure A.8. Autocorrelation as a function of measurement lag *l* for the 1.7-km time series shown in Fig. A.6. In the inset near the bottom of the figure, a zoom of the autocorrelation of reflectivity values between lag 0 and 5 is shown, together with a regression of the values between lag 2 and 5 extending toward lag 0 and plotted as a dashed line.

reminding us of the limitations of assuming an exponential correlation function as was done to estimate the correlation distance. Near zero lag, the autocorrelation often deviates from the exponential function it assumes at higher lag. If, like in this case, the autocorrelation function is convex, it suggests that some smoothing between nearby points occurred; this is the case here, because even if there was no overlap in time between measurements, the nonzero beam width caused some of the targets to be observed twice in successive measurements, resulting in an effective overlap of sampling volumes in space. If the autocorrelation is concave and shows a sharp peak near 0, it suggests that measurements are noisy; the contribution of the noise to the autocorrelation can then be estimated by using the difference between the point where the extrapolation of autocorrelation at higher lags intercepts lag 0 (the dashed line in Fig. A.8) and 1, the value expected for perfect correlation. This difference is known as the nugget, higher positive nuggets being associated with noisier or less representative data.

A.4.4 From sample to population statistics: challenges

In all the discussion above, I have tried to be careful and specified that all the statistics computed were for the samples we measured. To learn something useful about a phenomenon, we need to infer the properties of the population we are trying to study using such samples. The key difference between those two is that the sample is the fraction of the population that is measured, while the population is the entire set of events or phenomena we want to study. Properties of interest include the mean, or the expected value, of a population, as well as its standard deviation σ. The tools and foundations of statistics rest on the premise that we can use information derived from a sample to determine population properties, as well as estimate the uncertainty with which these can be determined. But geophysical fields with their unusual structure make such an inference difficult.

Geophysical fields in general have structures of different size imbedded in each other. For example, we previously mentioned how, for winds, large turbulent structures always have progressively smaller structures imbedded; in precipitation, we are familiar with cells that tend to be within clusters, themselves tending to be within atmospheric systems. If we try to decompose such patterns into their contribution from different scales, using for example a Fourier transform, we realize that patterns of increasing scale, or of decreasing wavenumber ω, have increasing amplitudes (the wavenumber ω is defined as $2\pi/L$, where L is the wavelength of the scale considered). If we characterize atmospheric fields using a power spectrum that measures the energy associated with patterns at each scale or wavenumber, we can often see that the power spectrum E_ω is proportional to ω^q, where the exponent q is a negative number. For winds, q is $-5/3$, while for precipitation at the mesoscale, q is about -1.4. Power-law energy spectra with negative exponents on the wavenumber are a common characteristic of fields that have correlation at all scales.

Fields with correlation at all scales are challenging to characterize. Most simple statistical tools assume that the events or measurements to be analyzed and combined are uncorrelated. To illustrate this fact, consider the five 2-D random fields shown in Fig. A.9, going from white noise (no correlation at any scale, $q = 0$) to Brownian noise

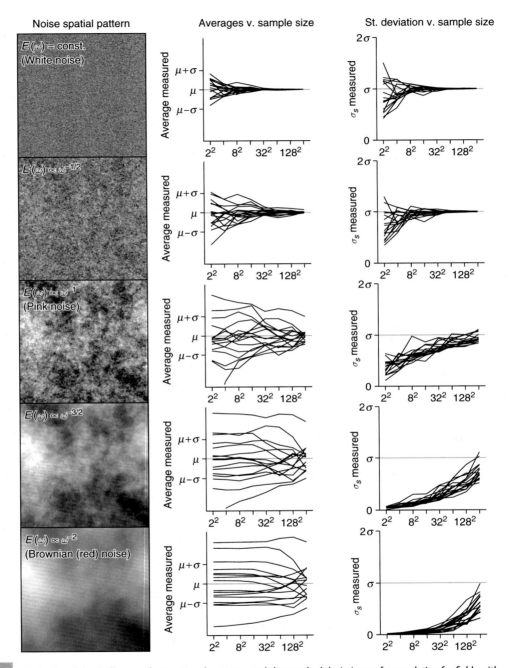

Figure A.9. Illustration of the challenges of estimating the mean μ and the standard deviation σ of a population for fields with correlated structures. On the left, five random 2-D fields with increasing variability at smaller wavenumber ω (or at larger scales) are shown. Attempts to estimate μ (middle) and σ (right) for 16 different samples whose size gradually increases are then illustrated.

($q = -2$). These examples span the kind of structures observed in many atmospheric fields; for example, the field with $q = -1.5$ has the look and feel of precipitation patterns observed on individual radars. All five fields of 1024×1024 points have by design an identical mean μ and standard deviation σ. As an exercise, we will try to use different samples within those fields and measure their average and sample standard deviation σ_s in an attempt to estimate μ and σ. The size of these samples will also be varied from 2^2 to 256^2, the expectation being that as sample size increases, our attempts at estimating μ and σ should become progressively better. The results of this simulation exercise are shown on the middle and right side of Fig. A.9.

If we first focus on the example of white noise, we can see that estimates of the mean and standard deviation are unbiased and rapidly improve as sample size increases. In fact, the uncertainty on the mean and the standard deviation is proportional to $\sigma/M_s^{-1/2}$, M_s being the number of measurements considered (from 2^2 to 256^2 in Fig. A.9). As q decreases though, two key phenomena happen: the rate of convergence toward the correct answer diminishes rapidly, and the estimates of standard deviations are biased low for small samples. When q reaches -1, standard deviations are biased low for all sample sizes; when q reaches -2, the sample averages fail to converge to the mean even as sample size increases. Hence, for geophysical fields in general and atmospheric phenomena at the mesoscale in particular, it can be extremely difficult, if not impossible, to estimate the mean and standard deviations of the properties of the fields observed.

This issue can be further illustrated by revisiting Figs. A.6–A.8. We saw that the lagged correlations in Figs. A.7 and A.8 were computed based on the time series of reflectivity plotted in Fig. A.6. But that time series is actually a small fraction of the one plotted in Fig. A.4. If we study Fig. A.4, we realize that the time series in Fig. A.6 corresponds to a relatively flat section between 23:05 and 23:25 of the larger time series. Were the time series to be extended to 22:00–24:00 for example, many results would change: not only would the mean and variance be different, but so would the average time of decorrelation, as it would now be primarily determined by the ~20-min ups and downs rather than the shorter fluctuations that gave us an answer of 40 s before. By slightly extending the data set used, the conclusions derived from it changed drastically.

Although this discussion may sound esoteric, its importance cannot be overstated. Our ability to define and generalize descriptions of atmospheric structures or patterns ultimately rests on our capacity to define means and standard deviations. If means and standard deviations cannot be computed, then efforts to describe patterns become largely futile. This partly explains why each atmospheric event appears so different yet somehow similar to another, and why the magnitude of case-to-case variability is so large that it challenges our ability to categorize events and generalize conclusions obtained from short-duration case studies. What this exercise also shows is that if the pattern of study has the kind of structures regularly observed in atmospheric fields, that is, small maxima within larger medium-size peaks within larger structures, then samples will on average underestimate the standard deviation of the population. Consequently, the likelihood of rare events that can be estimated from a sample is generally an underestimate with respect to reality. In parallel, the rate at which samples of increasing size converge to the correct answer is always slower than one expects assuming statistics based on uncorrelated events. This is a sobering result given, for example, the societal need to describe the likelihood of disruptive atmospheric and geophysical phenomena.

A.5 Basic radar signal processing

Once the signal coming from the receiver has been digitized, it must be processed in order to obtain estimates of radar measurements such as received power or mean Doppler velocity. The signal being digitized is the intermediate frequency signal illustrated in Fig. 13.4. All single-polarization radar measurements are derived from one such signal, typically the echo at horizontal polarization obtained after transmitting a wave at horizontal polarization. Dual-polarization measurements are made by combining two or more signals of such signals, such as HH (horizontal polarization on transmit and on receive), VV (vertical polarization on transmit and on receive), and sometimes VH (vertical polarization on transmit, horizontal polarization on receive). All these signals oscillate around a dominant frequency corresponding to the intermediate frequency f_{IF}.

For reasons that will only become clear later, it is much simpler and cleaner to make computations on oscillating signals using complex number algebra. As a result, all radar signal processing algorithms and the literature on the subject treat the signal as a time series of complex numbers. In order to explain signal processing, a quick refresher on complex numbers and some of their properties is hence required.

A.5.1 Simple operations on complex numbers

A complex number z has a real part a and an imaginary part b. Mathematically, it is expressed as $z = a + ib$, where $i = \sqrt{-1}$ (mathematicians and physical scientists generally use i to symbolize the square root of -1, while engineers use j; here I chose to use the mathematical convention). Graphically, complex numbers are often illustrated as a vector on a 2-D plane, with the abscissa corresponding to the real part and the ordinate corresponding to the imaginary part (Fig. A.10). A complex number $z = a + ib$ has

– a real part: $\mathrm{Re}\{z\} = a$,
– an imaginary part: $\mathrm{Im}\{z\} = b$, where b itself is a real number,
– a modulus: $|z| = \sqrt{a^2 + b^2}$, also a real number, and
– a complex conjugate: $z^* = a - ib$.

Arithmetic operations on complex numbers are similar to those on real numbers. Useful examples include the following:

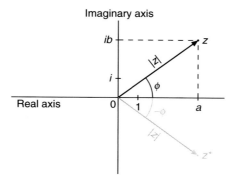

Graphical representation of a complex number z.

$$z_1 + z_2 = (a_1 + ib_1) + (a_2 + ib_2) = (a_1 + a_2) + i(b_1 + b_2)$$
$$i^2 = -1$$
$$z_1 z_2 = (a_1 + ib_1)(a_2 + ib_2) = (a_1 a_2 - b_1 b_2) + i(a_2 b_1 + a_1 b_2)$$
$$z_1 z_1^* = (a_1 + ib_1)(a_1 - ib_1) = a_1^2 + b_1^2 = |z_1|^2$$
$$\frac{z_1}{z_2} = \frac{(a_1 + ib_1)}{(a_2 + ib_2)} = \frac{(a_1 + ib_1)}{(a_2 + ib_2)} \frac{(a_2 - ib_2)}{(a_2 - ib_2)} = \frac{(a_1 a_2 + b_1 b_2) + i(a_2 b_1 - a_1 b_2)}{a_2^2 + b_2^2} = \frac{z_1 z_2^*}{|z_2|^2}.$$

$$(A.17)$$

In addition to the usual $z = a + ib$, there are two other very useful ways of expressing z especially in the context of oscillating signals. These are the polar form and the exponential form of complex numbers. The polar form expresses the complex number as a combination of the modulus $|z|$ and of an angle, the argument φ_z. Specifically,

$$z = |z| \frac{a + ib}{\sqrt{a^2 + b^2}} = |z|(\cos \varphi_z + i \sin \varphi_z),$$

$$\varphi_z = \arg(z) = \tan_2^{-1}(b, a)$$

$$(A.18)$$

where $\tan_2^{-1}(b, a)$ is the two-parameter version of arctangent such that

$$\tan_2^{-1}(b, a) = \begin{cases} \tan^{-1}\left(\dfrac{b}{a}\right) & \text{for } a > 0 \\[2mm] \pi + \tan^{-1}\left(\dfrac{b}{a}\right) & \text{for } a < 0 \\[2mm] \pi/2 \text{ for } a = 0 \text{ and } b > 0 \\[1mm] -\pi/2 \text{ for } a = 0 \text{ and } b < 0 \end{cases}.$$

$$(A.19)$$

The exponential form is rooted on the definition of the exponential function applied to complex numbers:

$$\exp(z) = \exp(a + ib) = \exp(a)\exp(ib) = \exp(a)(\cos b + i \sin b), \tag{A.20}$$

as a result of which,

$$z = |z|(\cos \varphi_z + i \sin \varphi_z) = |z|e^{i\varphi_z}. \tag{A.21}$$

The exponential form is very convenient for multiplying complex numbers and raising them to a given power, the properties of the exponential function for complex numbers being similar to those for real numbers.

A.5.2 Radar signal as a time series of complex numbers

The digitized raw signal coming from the receiver near the intermediate frequency (Fig. 13.4) is difficult to interpret directly: each individual measurement originates from a signal of varying amplitude that is multiplied by a near-sinusoidal function whose frequency centered on $f_{IF} = f - f_{LO}$ can fluctuate by the bandwidth of the receiver. From such a signal, it is not immediately obvious how the amplitude of the signal or especially its phase can be determined.

Given that the signal has nevertheless a known expected central frequency f_{IF}, we can take advantage of that knowledge to compute the phase shift of the signal with respect to a known reference signal of frequency f_{IF} (Fig. 13.4). Such a computation is done over a short interval that is a multiple M_λ of the wavelength of f_{IF}. Consider the radar signal and a known signal of frequency f_{IF} with unit amplitude. If we assume that the radar signal locally has a fixed frequency $f - f_{LO}$ and has an initial phase shift of φ with respect to the reference signal of frequency f_{IF}, the integral of the product between the radar signal and the known reference over M_λ wavelengths of f_{IF} is a quantity $M_\lambda cI/f_{IF}$ such that

$$\frac{M_\lambda c}{f_{IF}} I = \int_0^{M_\lambda c/f_{IF}} A \sin[2\pi(f - f_{LO})t + \varphi] \sin(2\pi f_{IF} t) dt$$

$$= \frac{A}{2} \left\{ \int_0^{M_\lambda c/f_{IF}} \cos[2\pi(f - f_{LO} - f_{IF})t + \varphi] - \cos[2\pi(f - f_{LO} + f_{IF})t + \varphi] dt \right\}$$

$$\approx \frac{A}{2} \left\{ \int_0^{M_\lambda c/f_{IF}} \cos(\varphi) - \cos(4\pi f_{IF} t + \varphi) dt \right\} = \frac{M_\lambda c}{f_{IF}} A \cos \varphi. \tag{A.22}$$

The approximation on the third line of (A.22) takes advantage of the fact that $f - f_{LO} \approx f_{IF}$. If we do the same type of calculation but with a reference frequency 90° out of phase compared with the one before, we obtain a quantity $M_\lambda cQ/f_{IF}$ such that

$$\frac{M_\lambda c}{f_{\mathrm{IF}}} Q = \int_0^{M_\lambda c/f_{\mathrm{IF}}} A \sin\left[2\pi(f - f_{\mathrm{LO}})t + \varphi\right] \sin\left(2\pi f_{\mathrm{IF}}t + \frac{\pi}{2}\right) dt$$

$$= \frac{A}{2}\left\{\int_0^{M_\lambda c/f_{\mathrm{IF}}} \sin\left[2\pi(f - f_{\mathrm{LO}} - f_{\mathrm{IF}})t + \varphi\right] + \sin\left[2\pi(f - f_{\mathrm{LO}} + f_{\mathrm{IF}})t + \varphi\right] dt\right\}$$

$$\approx \frac{M_\lambda c}{f_{\mathrm{IF}}} A \sin\varphi. \tag{A.23}$$

These derived quantities I and Q are hence the product of the amplitude of the signal and of the cosine and sine of the phase, respectively. It then becomes simple to use I and Q to determine the amplitude A and phase φ of the signal as a function of range and time. If we then choose to use a complex representation for (I, Q) such that $z = I + iQ$, then $|z| = A$ and $\arg(z) = \varphi$.

The (I, Q) or complex representation of radar signals has some nice properties. For example, if the signal is from a single target, $|z| = A$ is constant if the power illuminating that target is constant, and only $\arg(z)$ varies as a function of small changes in path properties (r and n). How the signals from multiple targets interfere with each other can be expressed as a sum of complex numbers or of 2-D vectors instead of a sum of oscillatory signals (Fig. A.11), the resulting modulus and argument being the final amplitude and phase of the combined signal.

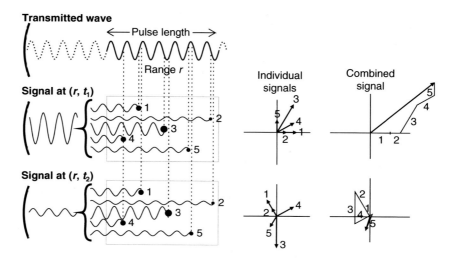

 Figure A.11. The interference of the signals from multiple targets illustrated with oscillatory signals as in Fig. 2.15 (left) compared with vectors or complex numbers on the right.

A.5.3 Covariances of time series of complex numbers

At a fixed range, given multiple pulses fired, a time series of (I, Q) is obtained for each polarization combination being digitized. Most radar measurements can then be derived by computing covariances on those time series.

The covariance at lag l between two complex data sets z_j and ζ_j is

$$\text{cov}_{z,\zeta}[l] = \begin{cases} \dfrac{1}{M_s - l - 1} \displaystyle\sum_{j=1}^{M_s - l} (z_j - \bar{z})^* (\zeta_{j+l} - \bar{\zeta}) & \text{for } l \geq 0 \\[2ex] \dfrac{1}{M_s + l - 1} \displaystyle\sum_{j=1-l}^{M_s} (z_{j-l} - \bar{z})^* (\zeta_j - \bar{\zeta}) & \text{for } l < 0 \end{cases}, \tag{A.24}$$

If targets are randomly distributed within the sampling volume, then all target phases between $-\pi$ and π are equiprobable. In parallel, noise has random phase. Given that all phases are equiprobable for both signal and noise, all arguments of z_i and ζ_i are also equiprobable. As a result, the mean of time series of radar signals is 0 and does not need to be estimated from the sample. Therefore, for radar signals, the covariance between two time series simplifies to

$$\text{cov}_{z,\zeta}[l] = \begin{cases} \dfrac{1}{M_s - l} \displaystyle\sum_{j=1}^{M_s - l} z_j^* \zeta_{j+l} & \text{for } l \geq 0 \\[2ex] \dfrac{1}{M_s + l} \displaystyle\sum_{j=1-l}^{M_s} z_{j-l}^* \zeta_j & \text{for } l < 0 \end{cases}. \tag{A.25}$$

Equation (A.25) is surprisingly powerful given its unassuming plainness. To illustrate its value, considering only positive lags for simplicity, we can express each complex number in its exponential form and obtain

$$\text{cov}_{z,\zeta}[l] = \frac{1}{M_s - l} \sum_{j=1}^{M_s - l} z_j^* \zeta_{j+l} = \frac{1}{M_s - l} \sum_{j=1}^{M_s - l} \left| z_j \zeta_{j+l} \right| \exp\left[i \left(\varphi_{\zeta_{j+l}} - \varphi_{z_j} \right) \right]. \tag{A.26}$$

Equation (A.26) suggests that covariances of complex radar signals that have amplitudes as their modulus and phases as their arguments will have a final modulus with units of power (or amplitude squared) and an argument that is a function of the average phase difference between the two time series.

Before applying (A.25) to radar time series, let us also define the lagged correlation coefficient for time series of complex numbers:

$$\rho_{z,\zeta}[l] = \frac{\text{cov}_{z,\zeta}[l]}{|\sigma_s(z)||\sigma_s(\varsigma)|} = \frac{|\text{cov}_{z,\zeta}[l]|}{|\sigma_s(z)||\sigma_s(\varsigma)|} \exp\{i \arg[\text{cov}_{z,\zeta}[l]]\}, \tag{A.27}$$

with

$$\sigma_s(z) = \sqrt{\frac{1}{M_s - l} \sum_{j=1}^{M_s - l} z_j^2} \quad \text{and} \quad \sigma_s(\zeta) = \sqrt{\frac{1}{M_s - l} \sum_{j=l+1}^{M_s} \zeta_j^2}, \qquad (A.28)$$

for positive l and given the fact that the mean value of both time series is 0. Correlation is therefore a complex number itself; its argument is that of the complex covariance while its modulus is the ratio of the modulus of the covariance and of the product of the modulus of the standard deviations of the two time series. In general though, when the term "correlation" is used in the context of radar time series, it refers to the modulus of the complex correlation only. A consequence of this definition of correlation is that while for real numbers, $-1 \leq \rho_{X,Y} \leq 1$, we have $0 \leq |\rho_{z,\zeta}| \leq 1$ for complex numbers. This small difference in the behavior of the correlation function will have an impact at low correlations: while two decorrelated time series of real numbers have on average a correlation of 0, with accidently correlated time series compensating other accidently anticorrelated ones, the average norm of the correlation of two independent complex time series is small but nonzero. This is important to remember when trying to compensate for the bias introduced by noise in estimates of correlation.

A.5.4 Radar measurables in the time domain

A.5.4.1 Autocovariances, reflectivity, and Doppler velocity

For single-polarization radars, only one time series of (I, Q) data is available at each range. Let us decompose each measurement $(I_j + i\, Q_j)$ into its signal component S_j and its noise component N_j. At first, let us consider the situation when signal is very strong and noise can be ignored. If we compute the autocovariances of lag 0 and 1 of a time series of (I, Q) dominated by signal, we obtain

$$\text{cov}_{S,S}[0] = \frac{1}{M_s} \sum_{j=1}^{M_s} S_j^* S_j = \overline{|S|^2} \qquad (A.29)$$

and

$$\text{cov}_{S,S}[1] = \frac{1}{M_s - 1} \sum_{j=1}^{M_s - 1} S_j^* S_{j+1} = \frac{1}{M_s - 1} \sum_{j=1}^{M_s - 1} |S_j S_{j+1}| \exp[i(\varphi_{S_{j+1}} - \varphi_{S_j})]. \qquad (A.30)$$

The lag 0 autocovariance of signal is a real number that corresponds to the average power of that signal and is proportional to the signal power P_r received by the radar. This signal power is then used to compute reflectivity using (3.2). The lag 1 autocovariance is a complex number whose argument is the average power-weighted phase shift between successive pulses. From the average phase shift between successive pulses with a pulse repetition frequency f_r, the mean radial velocity is derived using (5.7). Using (I, Q) data, (A.30) is computed using what is known as a pulse pair algorithm whereby

$$S_j^* S_{j+1} = (I_{S_j} - iQ_{S_j})(I_{S_{j+1}} + iQ_{S_{j+1}}) = (I_{S_j}I_{S_{j+1}} + Q_{S_j}Q_{S_{j+1}}) + i(I_{S_j}Q_{S_{j+1}} - Q_{S_j}I_{S_{j+1}}).$$

$$(A.31)$$

If we recall that $I = A\ \cos(\varphi)$ and $Q = A\ \sin(\varphi)$, we may recognize in (A.31) familiar relationships based on geometry: $\cos(X-Y) = \cos(X)\cos(Y) + \sin(X)\sin(Y)$ for the real part and $\sin(X-Y) = \sin(X)\cos(Y) - \cos(X)\sin(Y)$ for the imaginary part. Finally, based on (A.27), the ratio of $|\mathrm{cov}_{S,S}[1]|$ and $\mathrm{cov}_{S,S}[0]$ is the magnitude of the lag 1 autocorrelation of the signal. If we recall that signal decorrelation is caused by the relative movement of targets with respect to each other, then $|\mathrm{cov}_{S,S}[1]|/\mathrm{cov}_{S,S}[0]$ can be used to obtain the spectrum width of velocity σ_v via

$$\sigma_v = \frac{cf_r}{4\pi fn}\sqrt{-2\log\left[|\mathrm{cov}_{S,S}[1]|/\mathrm{cov}_{S,S}[0]\right]}, \qquad (A.32)$$

where f is the radar transmit frequency, f_r is the pulse repetition frequency, and n is the refractive index of air.

Complications arise when noise must be taken into account. In that case, the lag 0 and lag 1 autocovariances become

$$\mathrm{cov}_{S+N,S+N}[0] = \frac{1}{M_s}\sum_{j=1}^{M_s}(S_j^* + N_j^*)(S_j + N_j) = \overline{|S+N|^2}$$

$$= \overline{|S|^2} + \overline{|N|^2} + \mathrm{cov}_{S,N}[0] + \mathrm{cov}_{N,S}[0] \qquad (A.33)$$

and

$$\mathrm{cov}_{S+N,S+N}[1] = \frac{1}{M_s - 1}\sum_{j=1}^{M_s-1}(S_j^* + N_j^*)(S_{j+1} + N_{j+1})$$

$$= \mathrm{cov}_{S,S}[1] + \mathrm{cov}_{N,N}[1] + \mathrm{cov}_{S,N}[1] + \mathrm{cov}_{N,S}[1] \qquad (A.34)$$

The lag 0 autocovariance now measures the power from the sum of signal plus noise. A proper estimate of noise power is hence required if we are to properly quantify signal strength. Because noise and signal are decorrelated, the sample covariances between signal and noise as well as the autocovariance of noise for lags different from 0 are generally small. However, when signal is very weak, even with noise power perfectly known, the estimate of signal power will have an additional uncertainty due to the unknown sample covariance between noise and signal.

While lag 0 autocovariance is caused by signal and noise, the modulus of lag 1 autocorrelation is dominated by the autocovariance of signal only, except when $|S|\ll|N|$. Therefore, if signal decorrelation is expected to be small, the modulus of lag 1 autocovariance becomes a decent estimate of signal power and is occasionally referred to as coherent power. Most affected by noise is the estimate of spectrum width, as it now relies on the ratio of a lag 1 autocovariance dominated by signal and a lag 0 autocovariance that has signal and noise; for proper estimates of spectrum width, noise power must first be removed from the lag 0 autocovariance. In parallel, since small ratios of lag 1 to lag 0 autocovariances are generally associated with less accurate estimates, that ratio is often used as a signal quality index.

A.5.4.2 Covariances and dual-polarization estimates

With dual-polarization radars, at least two time series are available. Z_{dr} being a simple ratio of the power at horizontal and vertical polarization, it can be computed using ratios of power estimated at H and V either based on lag 0 or on lag 1 autocovariances. Other measured quantities are derived from the covariance of the signal received at H and V:

$$\text{cov}_{V,H}[l] = \frac{1}{M_s - l} \sum_{j=1}^{M_s - l} V_j^* H_{j+l} = \frac{1}{M_s - l} \sum_{j=1}^{M_s - l} |V_j H_{j+l}| \exp[i(\varphi_{H_{j+l}} - \varphi_{V_j})], \qquad (A.35)$$

where H_j and V_j refer to elements of the (I, Q) time series at horizontal and vertical polarization, respectively. At lag 0, the argument of the covariance is the mean phase shift between the time series at horizontal and vertical polarization and corresponds to ψ_{dp}, from which Φ_{DP} is derived using (6.1). As for ρ_{co}, it can be estimated using any lag l from

$$\rho_{co}[l] = \rho_{V,H}[l] = \frac{|\text{cov}_{V,H}[l]|}{\sqrt{|\text{cov}_{H,H}[l]||\text{cov}_{V,V}[l]|}}. \qquad (A.36)$$

Lag 1 estimates provide results that are the least sensitive to noise (though not entirely insensitive), an important characteristic given that a small contamination of ρ_{co} by noise may be sufficient to change ρ_{co} from a value expected for meteorological targets to one expected for nonmeteorological targets. However, for strong targets and where high spectrum widths are expected, lag 0 estimates are more precise.

A.5.5 Doppler spectra

As seen above, traditional time-domain approaches rely heavily on covariance calculations to estimate reflectivity, Doppler velocity, and spectrum width. Another possibility is to use what are known as frequency-domain approaches whereby time series of (I, Q) data are first decomposed in a sum of sinusoidal functions each having a different frequency.

A.5.5.1 Fourier series and transforms

The idea behind a Fourier series applied to complex numbers is that any (reasonable) complex function $f(X)$ that is periodic over an interval of width L can be decomposed into a sum of periodic functions multiplied by a weight:

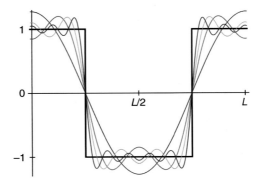

Figure A.12. A periodic square wave is decomposed using a varying number of Fourier terms as defined in (A.37): $j \leq 1$ (blue), $j \leq 3$ (green), $j \leq 5$ (yellow), and $j \leq 9$ (red).

$$f(X) = \sum_{j=-\infty}^{\infty} c_j \exp\left(i\frac{2\pi j X}{L}\right)$$

$$c_j = \frac{1}{L} \int_{-L/2}^{L/2} f(X) \exp\left(-i\frac{2\pi j X}{L}\right) dx$$

(A.37)

where j is an integer and c_j is a set of complex numbers representing the weights applied to the exponential basis functions. The functions $\exp(i2\pi j x/L)$ can be used for such a decomposition because they satisfy the following orthogonality conditions:

$$\int_{-\pi}^{\pi} \exp(ij_1 X)\exp(-ij_2 X)dX = \begin{cases} 0 & \text{for } j_1 \neq j_2 \\ 2\pi & \text{for } j_1 = j_2 \end{cases}.$$

(A.38)

A remarkable property of Fourier transforms is that with enough terms in c_j, any reasonable function can be approached, even those one would not expect could be reproduced by smooth basis functions (Fig. A.12).

The Fourier transform is a generalization of the Fourier series in the limit $L \to \infty$. When that occurs, there are so many basis functions that they become a continuous function. If one replaces the discrete c_j by a continuous $F(\omega)d\omega$ while letting $n/L \to \omega$, and change the sum into an integral, we get

$$f(X) \equiv \mathcal{F}^{-1}[F(\omega)] = \int_{-\infty}^{\infty} F(\omega)\exp(i2\pi\omega X)d\omega$$

$$F(\omega) \equiv \mathcal{F}[f(X)] = \int_{-\infty}^{\infty} f(X)\exp(-i2\pi\omega X)dX$$

(A.39)

Here $\mathcal{F}[f(X)]$ describes the forward Fourier transform that allows us to go from $f(X)$ to $F(\omega)$ while $\mathcal{F}^{-1}[F(\omega)]$ is the inverse Fourier transform (to go from $F(\omega)$ to $f(X)$). The Fourier transform represents a decomposition of a function in frequency space.

A.5.5.2 Discrete Fourier transform and power spectrum

If instead of having a continuous function $f(X)$ one has a function defined at a discrete number of points M_s such that $f_j = f(jL/M_s)$ with $j = 0, \ldots, M_s-1$, one can decompose that function into M_s discrete Fourier terms F_ω with $\omega = 0, \ldots, M_s-1$ (or $\omega = -M_s/2, \ldots, M_s/2-1$) such that

$$F_\omega = \sum_{j=0}^{M_s-1} f_j \exp\left(-i\frac{2\pi\omega j}{M_s}\right)$$

and

$$f_j = \frac{1}{M_s}\sum_{j=0}^{M_s-1} F_\omega \exp\left(i\frac{2\pi\omega j}{M_s}\right) \tag{A.40}$$

Discrete Fourier transforms are extremely useful because they reveal periodicities in input data as well as the relative strengths of any periodic components. To further look at the strength, or power, of these periodicities, one generally uses the power spectrum:

$$E_\omega = F_\omega F_\omega^*. \tag{A.41}$$

Power spectra are used in countless data analysis and signal processing tasks. For example, the complex radar signal can be decomposed into its Fourier components, from which a power spectrum can be determined. This power spectrum can then reveal what fraction of the signal power moves at what velocity (e.g., Fig. 5.3). From these distributions of power at different velocities, signal power, noise power, mean Doppler velocity, and spectrum width can be determined. Power spectra can also be used to identify and then filter unwanted ground echoes that can be recognized as an anomalous peak at zero Doppler velocity.

References

Alexander, C. R., S. S. Weygandt, T. G. Smirnova, *et al.*, 2010: High Resolution Rapid Refresh (HRRR): Recent enhancements and evaluation during the 2010 convective season. Preprints, *25th Conference on Severe Local Storms*, Denver, CO, October 11–14, 2010, American Meteorological Society, **9**.2.

American Meteorological Society, cited 2014: Baroclinic instability. Glossary of meteorology. [Available online at: http://glossary.ametsoc.org/wiki/Baroclinic_instability].

Anderson, J. L., T. Hoar, K. Raeder, *et al.*, 2009: The Data Assimilation Research Testbed: A community facility. *Bulletin of the American Meteorological Society*, **90**, 1283–1296.

Angevine, W. M., A. W. Grimsdell, L. M. Hartten, and A. C. Delany, 1998: The Flatland Boundary Layer Experiments. *Bulletin of the American Meteorological Society*, **79**, 419–431.

Atlas, D., 1955: The origin of "stalactites" in precipitation echoes. *Proceedings of the Fifth Weather Radar Conference,* U.S. Signal Corps Engineering Laboratory, Fort Monmouth, NJ, September 12–15, 1955, American Meteorological Society, 321–328.

Atlas, D. (ed.), 1990: *Radar in Meteorology.* Boston, MA, American Meteorological Society, 806 pp.

Atlas, D., and C. W. Ulbrich, 1977: Path- and area-integrated rainfall measurement by microwave attenuation in the 1–3cm band. *Journal of Applied Meteorology*, **16**, 1322–1331.

Bally, J., 2004: The Thunderstorm Interactive Forecast System: Turning automated thunderstorm tracks into severe weather warnings. *Weather Forecasting*, **19**, 64–72.

Bannister, R. N., 2008: A review of forecast error covariance statistics in atmospheric variational data assimilation. II: Modelling the forecast error covariance statistics. *Quarterly Journal of the Royal Meteorological Society*, **134**, 1971–1996.

Bean, B. R., and E. J. Dutton, 1966: *Radio Meteorology.* National Bureau of Standards Monograph #92, U.S. Government Printing Office, 435 pp.

Bellon, A., and G. L. Austin, 1978: The evaluation of two years of a real-time operation of a short-term precipitation forecasting procedure (SHARP). *Journal of Applied Meteorology*, **17**, 1778–1787.

Bellon, A., and F. Fabry, 2014: Real-time radar reflectivity calibration from differential phase measurements. *Journal of Atmospheric and Oceanic Technology*, **31**, 1089–1097.

Berenguer, M., and I. Zawadzki, 2008: A study of the error covariance matrix of radar rainfall estimates in stratiform rain. *Weather and Forecasting*, **23**, 1085–1101.

Berenguer, M., D. Sempere-Torres, and G. G. S. Pegram, 2011: SBMcast – An ensemble nowcasting technique to assess the uncertainty in rainfall forecasts by Lagrangian extrapolation. *Journal of Hydrology*, **404**, 226–240.

Berne, A., and W. F. Krajewski, 2013: Radar for hydrology: Unfulfilled promise or unrecognized potential? *Advances in Water Resources*, **51**, 357–366.

Bowler, N. E., C. E. Pierce, and A. W. Seed, 2004: Development of a precipitation nowcasting algorithm based upon optical flow techniques. *Journal of Hydrology*, **288**, 74–91.

Brandes, E. A., and K. Ikeda, 2004: Freezing-level estimation with polarimetric radar. *Journal of Applied Meteorology*, **43**, 1541–1553.

Bringi, V. N., and V. Chandrasekar, 2001: *Polarimetric Doppler Weather Radar.* Cambridge, Cambridge University Press, 636 pp.

Brown, R. A., and V. T. Wood, 2006: *A Guide for Interpreting Doppler Velocity Patterns: Northern Hemisphere Edition.* Published by the National Severe Storms Laboratory. [Available at: www.nssl.noaa.gov/publications/dopplerguide/Doppler%20Guide%202nd%20Ed.pdf].

Chandrasekar, V., and S. Lim, 2008: Retrieval of reflectivity in a networked radar environment. *Journal of Atmospheric and Oceanic Technology*, **25**, 1755–1767.

Chandrasekar, V., R. Meneghini, and I. Zawadzki, 2003: Global and local precipitation measurements by radar. In *Radar and Atmospheric Science: A Collection of Essays in Honor of David Atlas*, R. M. Wakimoto and R. C. Srivastava (eds.). American Meteorological Society Monograph #52, 215–236.

Chilson, P. B., W. F. Frick, J. F. Kelly, *et al.*, 2012: Partly cloudy with a chance of migration – Weather, radars, and aeroecology. *Bulletin of the American Meteorological Society*, **93**, 669–686.

Clark, R. A., and D. R. Greene, 1972: Vertically integrated liquid water – A new analysis tool. *Monthly Weather Review*, **100**, 548–552.

Cohn, S. A., W. O. J. Brown, C. L. Martin, *et al.*, 2001: Clear air boundary layer spaced antenna wind measurement with the Multiple Antenna Profiler (MAPR). *Annals of Geophysics*, **19**, 845–854.

de Elía, R., and I. Zawadzki, 2000: Sidelobe contamination in bistatic radars. *Journal of Atmospheric and Oceanic Technology*, **17**, 1313–1329.

Delrieu, G., S. Caoudal, and J. D. Creutin, 1997: Feasibility of using mountain return for the correction of ground-based X-band weather radar data. *Journal of Atmospheric and Oceanic Technology*, **14**, 367–385.

Delrieu, G., A. Wijbrans, B. Boudevillain, *et al.*, 2014: Geostatistical radar–raingauge merging: A novel method for the quantification of rain estimation accuracy. *Advances in Water Resources*, **71**, 110–124.

Dixon, M., and G. Weiner, 1993: TITAN, Thunderstorm Identification, Tracking, Analysis and Nowcasting – A radar-based methodology. *Journal of Atmospheric and Oceanic Technology*, **10**, 785–797.

Donaldson, N., 2012: Interaction between beam blockage and vertical reflectivity gradients. *Proceedings of the Seventh European Conference on Radar in Meteorology and Hydrology*, Toulouse, France, June 24–29, 2012, Paper 8A.1, 5 pp. [Available at: http://www.meteo.fr/cic/meetings/2012/ERAD/extended_abs/DQ_080_ext_abs.pdf].

Ellis, S. M., and J. Vivekanandan, 2010: Water vapor estimates using simultaneous dual-wavelength radar observations. *Radio Science*, **45**, RS5002, doi:10.1029/2009RS004280.

Fabry, F., 1993: Wind profile estimation by conventional radars. *Journal of Applied Meteorology*, **32**, 40–49.

Fabry, F., 2006: The spatial structure of moisture near the surface: Project-long characterization. *Monthly Weather Review*, **134**, 79–91.

Fabry, F., and R. J. Keeler, 2003: Innovative signal utilization and processing. In *Radar and Atmospheric Science: A Collection of Essays in Honor of David Atlas*, R. M. Wakimoto and R. C. Srivastava (eds.). American Meteorological Society Monograph #52, 199–214.

Fabry, F., and A. Seed, 2009: Quantifying and predicting the accuracy of radar-based quantitative precipitation forecasts. *Advances in Water Resources*, **32**, 1043–1049.

Fabry, F., and I. Zawadzki, 1995: Long-term radar observations of the melting layer of precipitation and their interpretation. *Journal of the Atmospheric Sciences*, **52**, 838–851.

Fabry, F., and I. Zawadzki, 2001: New observational technologies: Scientific and societal impacts. In *Meteorology at the Millennium*, R. B. Pearce (ed.). London, UK, Academic Press, 72–82.

Fabry, F., G. L. Austin, and D. Tees, 1992: The accuracy of rainfall estimates by radar as a function of range. *Quarterly Journal of the Royal Meteorological Society*, **118**, 435–453.

Fabry, F., I. Zawadzki, and S. Cohn, 1993: The influence of stratiform precipitation on shallow convective rain: A case study. *Monthly Weather Review*, **121**, 3312–3325.

Fabry, F., A. Bellon, M. R. Duncan, and G. L. Austin, 1994: High resolution rainfall measurement by radar for very small basins: The sampling problem reexamined. *Journal of Hydrology*, **161**, 415–428.

Fabry, F., B. J. Turner, and S. A. Cohn, 1995: The University of Wyoming King Air educational initiative at McGill. *Bulletin of the American Meteorological Society*, **76**, 1806–1811.

Fabry, F., C. Frush, I. Zawadzki, and A. Kilambi, 1997: On the extraction of near-surface index of refraction using radar phase measurements from ground targets. *Journal of Atmospheric and Oceanic Technology*, **14**, 978–987.

Fleming, J. R. (ed.), 1996: *Historical Essays on Meteorology 1919–1995*. Boston, MA, American Meteorological Society, 618 pp.

Frisch, A. S., C. W. Fairall, and J. B. Snyder, 1995: Measurement of stratus cloud and drizzle parameters in ASTEX with a K-α-band Doppler radar and a microwave radiometer. *Journal of the Atmospheric Sciences*, **52**, 2788–2799.

Frush, C., R. J. Doviak, M. Sachidananda, and D. S. Zrnić, 2002: Application of the SZ phase code to mitigate range–velocity ambiguities in weather radars. *Journal of Atmospheric and Oceanic Technology*, **19**, 413–430.

Fulton, R. A., J. P. Breidenbach, D.-J. Seo, D. A. Miller, and T. O'Bannon, 1998: The WSR-88D rainfall algorithm. *Weather and Forecasting*, **13**, 377–395.

Gao, J., M. Xue, K. Brewster, and K. K. Droegemeier, 2004: A three-dimensional variational data analysis method with recursive filter for Doppler radars. *Journal of Atmospheric and Oceanic Technology*, **21**, 457–469.

Ge, G., J. Gao, K. Brewster, and M. Xue, 2010: Impacts of beam broadening and Earth curvature on storm-scale 3D variational data assimilation of radial velocity with two Doppler radars. *Journal of Atmospheric and Oceanic Technology*, **27**, 617–636.

Geerts, B., and Q. Miao, 2005: The use of millimeter Doppler radar echoes to estimate vertical air velocities in the fair-weather convective boundary layer. *Journal of Atmospheric and Oceanic Technology*, **22**, 225–246.

Germann, U., and I. Zawadzki, 2002: Scale-dependence of the predictability of precipitation from continental radar images. Part I: Description of the methodology. *Monthly Weather Review*, **130**, 2859–2873.

Germann, U., G. Galli, M. Boscacci, and M. Bolliger, 2006: Radar precipitation measurement in a mountainous region. *Quarterly Journal of the Royal Meteorological Society*, **132**, 1669–1692.

Giangrande, S. E., E. P. Luke, and P. Kollias, 2012: Characterization of vertical velocity and drop size distribution parameters in widespread precipitation at ARM facilities. *Journal of Applied Meteorology and Climatology*, **51**, 380–391.

Goddard, J. W. F., 1994: Technique for calibration of meteorological radars using differential phase. *Electronics Letters*, **30**, 166–167.

Gunn, R., and G. D. Kinzer, 1949: The terminal velocity of fall for water droplets in stagnant air. *Journal of Meteorology*, **6**, 243–248.

Habib, E., and W. F. Krajewski, 2002: Uncertainty analysis of the TRMM ground-validation radar-rainfall products: Application to the TEFLUN-B field campaign. *Journal of Applied Meteorology*, **41**, 558–572.

Harrison, D. L., K. Norman, C. Pierce, and N. Gaussiat, 2012: Radar products for hydrological applications in the UK. *Proceedings of the ICE – Water Management*, **165**, 89–103.

Hitschfeld, W., and J. Bordan, 1954: Errors inherent in the radar measurement of rainfall at attenuating wavelengths. *Journal of Meteorology*, **11**, 58–67.

Hocking, W. K., 2011: A review of Mesosphere–Stratosphere–Troposphere (MST) radar developments and studies, circa 1997–2008. *Journal of Atmospheric and Solar-Terrestrial Physics*, **73**, 848–882.

Hogan, R. J., C. Jakob, and A. J. Illingworth, 2001: Comparison of ECMWF winter-season cloud fraction with radar-derived values. *Journal of Applied Meteorology*, **40**, 513–525.

Houze, R. A., and P. V. Hobbs, 1982: Organization and structure of precipitating cloud systems. *Advances in Geophysics*, **24**, 225–315.

Huuskonen, A., E. Saltikoff, and I. Holleman, 2014: The operational weather radar network in Europe. *Bulletin of the American Meteorological Society*, **95**, 897–907.

Illingworth, A. J., R. J. Hogan, E. J. O'Connor, *et al.*, 2007: Cloudnet. *Bulletin of the American Meteorological Society*, **88**, 883–898.

Joe, P., S. Dance, V. Lakshmanan, *et al.*, 2012: Automated processing of Doppler radar data for severe weather warnings. In *Doppler Radar Observations – Weather Radar, Wind Profiler, Ionospheric Radar, and Other Advanced Applications*, J. Bech and J. L. Chau (eds.). Rijeka, Croatia, InTech, 33–74.

Johnson, J. T., P. L. MacKeen, A. Witt, *et al.*, 1998: The storm cell identification and tracking algorithm: An enhanced WSR-88D algorithm. *Weather and Forecasting*, **13**, 263–276.

Jorgensen, D. P., and T. M. Weckwerth, 2003: Forcing and organization of convective systems. In *Radar and Atmospheric Science: A Collection of Essays in Honor of David*

Atlas, R. M. Wakimoto and R. C. Srivastava (eds.). Boston, MA, American Meteorological Society, 75–103.

Jorgensen, D. P., T. Matejka, and J. D. DuGranrut, 1996: Multi-beam techniques for deriving wind fields from airborne Doppler radars. *Meteorology and Atmospheric Physics*, **59**, 83–104.

Joss, J., and A. Waldvogel, 1970: Raindrop size distributions and Doppler velocities. Preprints, *14th Radar Meteorology Conference*, Tucson, AZ, American Meteorological Society, 153–156.

Joss, J., and A. Waldvogel, 1990: Precipitation measurement and hydrology. In *Radar in Meteorology*, D. Atlas (ed.). Boston, MA, American Meteorological Society, 577–606.

Kay, J. E., and A. Gettelman, 2009: Cloud influence on and response to seasonal Arctic sea ice loss. *Journal of Geophysical Research*, **114**, D18204, doi:10.1029/2009JD011773.

Knight, C. A., and L. J. Miller, 1998: Early radar echoes from small, warm cumulus: Bragg and hydrometeor scattering. *Journal of the Atmospheric Sciences*, **55**, 2974–2992.

Kollias, P., B. A. Albrecht, R. Lhermitte, and A. Savtchenko, 2001: Radar observations of updrafts, downdrafts, and turbulence in fair-weather cumuli. *Journal of the Atmospheric Sciences*, **58**, 1750–1766.

Kollias, P., E. E. Clothiaux, M. A. Miller, *et al.*, 2007: Millimeter-wavelength radars: new frontier in atmospheric cloud and precipitation research. *Bulletin of the American Meteorological Society*, **88**, 1608–1624.

Kucera, P. A., W. F. Krajewski, and C. B. Young, 2004: Radar beam occultation studies using GIS and DEM technology: An example study. *Journal of Atmospheric and Oceanic Technology*, **21**, 995–1006.

Kumjian, M. R., 2013a: Principles and applications of dual-polarization weather radar. Part II: Warm- and cold-season applications. *Journal of Operational Meteorology*, **1**, 243–264.

Kumjian, M. R., 2013b: Principles and applications of dual-polarization weather radar. Part III: Artifacts. *Journal of Operational Meteorology*, **1**, 265–274.

Kumjian, M. R., A. V. Ryzhkov, H. D. Reeves, and T. J. Schuur, 2013: A dual-polarization radar signature of hydrometeor refreezing in winter storms. *Journal of Applied Meteorology and Climatology*, **52**, 2549–2566.

Lauri, T., J. Koistinen, and D. Moisseev, 2012: Advection-based adjustment of radar measurements. *Monthly Weather Review*, **140**, 1014–1022.

Lazo, J. K., R. E. Morss, and J. L. Demuth, 2009: 300 billion served: Sources, perceptions, uses, and values of weather forecasts. *Bulletin of the American Meteorological Society*, **90**, 785–798.

Lee, G. W., and I. Zawadzki, 2005a: Variability of drop size distributions: Time-scale dependence of the variability and its effects on rain estimation. *Journal of Applied Meteorology*, **44**, 241–255.

Lee, G. W., and I. Zawadzki, 2005b: Variability of drop size distributions: Noise and noise filtering in disdrometric data. *Journal of Applied Meteorology*, **44**, 634–652.

Lee, G. W., and I. Zawadzki 2006: Radar calibration by gage, disdrometer, and polarimetry: Theoretical limit caused by the variability of drop size distribution and application to fast scanning operational radar data. *Journal of Hydrology*, **328**, 83–97.

Lee, W.-C., F. D. Marks, and C. Walther, 2003: Airborne Doppler radar data analysis workshop. *Bulletin of the American Meteorological Society*, **84**, 1063–1075.

Lemon, L. R., 1980: Severe thunderstorm radar identification techniques and warning criteria. *NOAA Technical Memorandum NWS NSSFC-3*, NOAA National Severe Storms Forecast Center, Kansas City, MO.

Lewis, J. M., S. Lakshmivarahan, and S. Dhall, 2006: *Dynamic Data Assimilation: A Least Squares Approach*. Cambridge, UK, Cambridge Press, 680 pp.

Lhermitte, R. M., 1988: Observations of rain at vertical incidence with a 94 GHz Doppler radar: An insight of Mie scattering. *Geophysical Research Letters*, **15**, 1125–1128.

Li, J., and K. Nakamura, 2002: Characteristics of the mirror image of precipitation observed by the TRMM precipitation radar. *Journal of Atmospheric and Oceanic Technology*, **19**, 145–158.

Luke, E. P., and P. Kollias, 2013: Separating cloud and drizzle radar moments during precipitation onset using Doppler spectra. *Journal of Atmospheric and Oceanic Technology*, **30**, 1656–1671.

Mahrt, L., and D. Vickers, 2005: Boundary-layer adjustment over small-scale changes of surface heat flux. *Boundary Layer Meteorology*, **116**, 313–330.

Markowski, P., and Y. Richardson, 2010: *Mesoscale Meteorology in Midlatitudes*. Chichester, UK, Wiley, 430 pp.

Markowski, P., Y. Richardson, J. Marquis, *et al.*, 2012: The pretornadic phase of the Goshen County, Wyoming, supercell of 5 June 2009 intercepted by VORTEX2. Part I: Evolution of kinematic and surface thermodynamic fields. *Monthly Weather Review*, **140**, 2887–2915.

Marshall, J. S., 1953: Precipitation trajectories and patterns. *Journal of Meteorology*, **10**, 25–29.

Marshall, J. S., and W. Hitschfeld, 1953: Interpretation of the fluctuating echo from randomly distributed scatterers. Part 1. *Canadian Journal of Physics*, **31**, 962–994.

Marshall, J. S., and W. McK. Palmer, 1948: The distribution of raindrops with size. *Journal of Meteorology*, **5**, 165–166.

Marshall, J. S., R. C. Langille, and W. McK. Palmer, 1947: Measurement of rainfall by radar. *Journal of Meteorology*, **4**, 186–192.

Melnikov, V., and S. Matrosov, 2013: Radar measurements of the axis ratios of cloud particles. Proceedings, *36th Conference on Radar Meteorology*, Breckenridge, CO, September 16–20, 2013, American Meteorological Society.

Meneghini, R., T. Iguchi, T. Kozu, *et al.*, 2000: Use of the surface reference technique for path attenuation estimates from TRMM precipitation radar. *Journal of Applied Meteorology*, **39**, 2053–2070.

Mittermaier, M. P., R. J. Hogan, and A. J. Illingworth, 2004: Using mesoscale model winds for correcting wind-drift errors in radar estimates of surface rainfall. *Quarterly Journal of the Royal Meteorological Society*, **130**, 2105–2123.

Mohr, C. G., and L. J. Miller, 1983: CEDRIC – A software package for Cartesian space editing, synthesis, and display of radar fields under interactive control. Preprints, *21st Conference on Radar Meteorology*. Edmonton, AB, Canada, September 19–23, 1983, American Meteorological Society, 569–574.

Mueller, C., T. Saxen, R. Roberts, *et al.*, 2003: NCAR auto-nowcast system. *Weather Forecasting*, **18**, 545–561.

Nesbitt, S. W., and A. M. Anders, 2009: Very high resolution precipitation climatologies from the Tropical Rainfall Measuring Mission precipitation radar. *Geophysical Research Letters*, **36**, doi:10.1029/2009GL038026.

Nicol, J. C., A. J. Illingworth, T. Darlington, and M. Kitchen, 2013: Quantifying errors due to frequency changes and target location uncertainty for radar refractivity retrievals. *Journal of Atmospheric and Oceanic Technology*, **30**, 2006–2024.

Parent du Châtelet, J., C. Boudjabi, L. Besson, and O. Caumont, 2012: Errors caused by long-term drifts of magnetron frequencies for refractivity measurement with a radar: Theoretical formulation and initial validation. *Journal of Atmospheric and Oceanic Technology*, **29**, 1428–1434.

Park, H. S., A. V. Ryzhkov, D. S. Zrnić, K.-E. Kim, 2009: The hydrometeor classification algorithm for the polarimetric WSR-88D: Description and application to an MCS. *Weather Forecasting*, **24**, 730–748.

Pierce, C., A. Seed, S. Ballard, D. Simonin, and Z. Li, 2012: Nowcasting. In *Doppler Radar Observations – Weather Radar, Wind Profiler, Ionospheric Radar, and Other Advanced Applications*, J. Bech and J. L. Chau (eds.). Rijeka, Croatia, InTech, 97–142.

Politovitch, M. K., and B. C. Bernstein, 1995: Production and depletion of supercooled liquid water in a Colorado winter storm. *Journal of Applied Meteorology*, **34**, 2631–2648.

Pruppacher, H. R., and K. V. Beard, 1970: A wind tunnel investigation of the internal circulation and shape of water drops falling at terminal velocity in air. *Quarterly Journal of the Royal Meteorological Society*, **96**, 247–256.

Radhakrishna, B., I. Zawadzki, and F. Fabry, 2012: Predictability of precipitation from continental radar images. Part V: Growth and decay. *Journal of the Atmospheric Sciences*, **69**, 3336–3349.

Roberts, R. D., F. Fabry, P. C. Kennedy, *et al.*, 2008: REFRACTT-2006: Real-time retrieval of high-resolution, low-level moisture fields from operational NEXRAD and research radars. *Bulletin of the American Meteorological Society*, **89**, 1535–1548.

Roberts, R. D., A. R. S. Anderson, E. Nelson, *et al.*, 2012: Impacts of forecaster involvement on convective storm initiation and evolution nowcasting. *Weather and Forecasting*, **27**, 1061–1089.

Rosenfeld, D., and C. W. Ulbrich, 2003: Cloud microphysical properties, processes, and rainfall estimation opportunities. In *Radar and Atmospheric Science: A Collection of Essays in Honor of David Atlas*, R. M. Wakimoto and R. C. Srivastava (eds.). Boston, MA, American Meteorological Society, 270 pp.

Rotunno, R., J. B. Klemp, and M. L. Weisman, 1988: A theory for strong, long-lived squall lines. *Journal of the Atmospheric Sciences*, **45**, 463–485.

Ryzhkov, A. V., T. J. Schuur, D. W. Burgess, *et al.*, 2005: The joint polarization experiment: Polarimetric rainfall measurements and hydrometeor classification. *Bulletin of the American Meteorological Society*, **86**, 809–824.

Ryzhkov, A., M. Diederich, P. Zhang, and C. Simmer, 2014: Potential utilization of specific attenuation for rainfall estimation, mitigation of partial beam blockage, and radar networking. *Journal of Atmospheric and Oceanic Technology*, **31**, 599–619.

Sachidananda, M., and D. S. Zrnić, 1986: Differential propagation phase shift and rainfall rate estimation. *Radio Science*, **21**, 235–247.

Sachidananda, M., and D. Zrnić, 1999: Systematic phase codes for resolving range overlaid signals in a Doppler weather radar. *Journal of Atmospheric and Oceanic Technology*, **16**, 1351–1363.

Saltikoff, E., H. Hohti, and P. Lopez, 2014: Some challenges of QPE in snow. Proceedings, *the Eighth European Conference on Radar in Meteorology and Hydrology*, Garmisch-Partenkirchen, Germany, September 1–5, 2014. [Available at: http://www.pa.op.dlr.de/erad2014/programme/ExtendedAbstracts/036_Saltikoff.pdf].

Seed, A. W., 2003: A dynamic and spatial scaling approach to advection forecasting. *Journal of Applied Meteorology*, **42**, 381–388.

Seliga, T. A., and V. N. Bringi, 1976: Potential use of radar differential reflectivity measurements at orthogonal polarizations for measuring precipitation. *Journal of Applied Meteorology*, **15**, 69–76.

Sirmans, D., D. S. Zrnić, and B. Bumgarner, 1976: Extension of maximum unambiguous Doppler velocity by use of two sampling rates. Preprints, *17th Conference on Radar Meteorology*, Seattle WA, October 26–29, 1976, American Meteorological Society, 23–28.

Skolnik, M., 2008: *Radar Handbook*, 3rd edn. New York NY, McGraw Hill, 1328 pp.

Steiner, M., R. A. Houze Jr., and S. E. Yuter, 1995: Climatological characterization of three-dimensional storm structure from operational radar and rain gauge data. *Journal of Applied Meteorology*, **34**, 1978–2007.

Stimson, G. W., 1998: *Introduction to Airborne Radar*, 2nd edn. Mendham, NJ, Scitech Publishing, 576 pp.

Stumpf, G. J., A. Witt, E. D. Mitchell, *et al.*, 1998: The National Severe Storms Laboratory mesocyclone detection algorithm for the WSR-88D. *Weather and Forecasting*, **13**, 304–326.

Sun, J., and N. A. Crook, 1997: Dynamical and microphysical retrieval from Doppler radar observations using a cloud model and its adjoint. Part I: Model development and simulated data experiments. *Journal of Atmospheric Sciences*, **54**, 1642–1661.

Sun, J., and J. W. Wilson, 2003: The assimilation of radar data for weather prediction. In *Radar and Atmospheric Science: A Collection of Essays in Honor of David Atlas*, R. M. Wakimoto and R. C. Srivastava (eds.). Boston, MA, American Meteorological Society, 175–198.

Tabary, P., 2007: The new French operational radar rainfall product. Part I: Methodology. *Weather and Forecasting*, **22**, 393–408.

Tatarskii, V. I, 1971: *The Effects of the Turbulent Atmosphere on Wave Propagation* (translated from Russian by the Israel Program for Scientific Translations Ltd, ISBN 0 7065 0680 4), Reproduced by National Technical Information Service, US Department of Commerce, Springfield, VA.

Testud, J., E. Le Bouar, E. Obligis, and M. Ali-Mehenni, 2000: The rain profiling algorithm applied to polarimetric weather radar. *Journal of Atmospheric and Oceanic Technology*, **17**, 332–356.

Thompson, R. J., A. J. Illingworth, and J. Ovens, 2011: Emission: A simple technique to correct rainfall estimates from attenuation due to both radome and heavy rainfall. Proceedings, *8th International Symposium Weather Radar and Hydrology*, April 18–21, 2011, Exeter, UK.

Thomspon, T. E., L. J. Wicker, and X. Wang, 2012: Impact from a volumetric radar-sampling operator for radial velocity observations within EnKF supercell assimilation. *Journal of Atmospheric and Oceanic Technology*, **29**, 1417–1427.

Torres, S., and C. Curtis, 2011: A fresh look at the range weighting function for modern weather radars. Proceedings, *35th Radar Conference on Radar Meteorology*, Pittsburgh, PA, September 25–30, 2011, American Meteorological Society. [Available at: https://ams.confex.com/ams/35Radar/webprogram/Manuscript/Paper191117/Radar %20Conference%202011.pdf].

Trapp, R. J., 2013: *Mesoscale-Convective Processes in the Atmosphere*. Cambridge, Cambridge University Press, 377 pp.

Tsonis, A. A, and G. L. Austin, 1981: An evaluation of extrapolation techniques for the short-term prediction of rain amounts. *Atmosphere–Ocean*, **19**, 54–65.

Turner, B. J., I. Zawadzki, and U. Germann, 2004: Predictability of precipitation from continental radar images. Part III: Operational nowcasting implementation (MAPLE). *Journal of Applied Meteorology*, **43**, 231–248.

Valdez, M. P., and K. C. Young, 1985: Number fluxes in equilibrium raindrop populations: A Markov chain analysis. *Journal of the Atmospheric Sciences*, **42**, 1024–1036.

Wakimoto, R. M., H. Murphey, R. Fovell, and W.-C. Lee, 2004: Mantle echoes associated with deep convection: Observations and numerical simulations. *Monthly Weather Review*, **132**, 1701–1720.

WDTB (Weather Decision Training Branch), 2014: *Distance Learning Operations Course Topic 7: Convective Storm Structure and Evolution*. [Available online at: http://www. wdtb.noaa.gov/courses/dloc/documentation/DLOC_FY14_Topic7.pdf, and try replacing "14" by the current year].

Weckwerth, T. M., C. R. Pettet, F. Fabry, S. Park, J. W. Wilson, and M. A. LeMone, 2005: Radar refractivity retrieval: Validation and application to short-term forecasting. *Journal of Applied Meteorology*, **44**, 285–300.

Williams, E., 2014: *Aviation Formulary V1.46*. [Available online at: http://williams.best. vwh.net/avform.htm].

Wilson, J. W., and E. A. Brandes, 1979: Radar measurement of rainfall – A summary. *Bulletin of the American Meteorological Society*, **60**, 1048–1058.

Wilson, J. W., and W. E. Schreiber, 1986: Initiation of convective storms at radar-observed boundary-layer convergence lines. *Monthly Weather Review*, **114**, 2516–2536.

Wurman, J., 1994: Vector winds from a single-transmitter bistatic dual-Doppler radar network. *Bulletin of the American Meteorological Society*, **75**, 983–994.

Xue, M., F. Kong, K. W. Thomas, *et al.*, 2008: CAPS realtime storm-scale ensemble and high-resolution forecasts as part of the NOAA Hazardous Weather Testbed 2008 Spring Experiment. Preprints, *24th Conference on Severe Local Storms*, Savannah, GA, October 27–31, 2008, American Meteorological Society, 12.2. [Available online at https://ams. confex.com/ams/24SLS/techprogram/paper_142036.htm].

Zawadzki, I., and M. De Agostinho Antonio, 1988: Equilibrium raindrop size distributions in tropical rain. *Journal of the Atmospheric Sciences*, **45**, 3452–3459.

Zawadzki, I., W. Szyrmer, and S. Laroche, 2000: Diagnostic of supercooled clouds from single-Doppler observations in regions of radar-detectable snow. *Journal of Applied Meteorology*, **39**, 1041–1058.

Zeng, Y., U. Blahak, M. Neuper, and D. Jerger, 2014: Radar beam tracing methods based on atmospheric refractive index. *Journal of Atmospheric and Oceanic Technology*, **31**, 2650–2670.

Zhang, G., and R. J. Doviak, 2007: Spaced-antenna interferometry to measure crossbeam wind, shear, and turbulence: Theory and formulation. *Journal of Atmospheric and Oceanic Technology*, **25**, 791–805.

Index